COMPUTER PROGRAMS
For Electronic Analysis and Design

COMPUTER PROGRAMS
For Electronic Analysis and Design

DIMITRI SPARTACO BUGNOLO

Associate Professor of
Electrical and Computer Engineering
Florida Institute of Technology
Melbourne, Florida

RESTON PUBLISHING COMPANY, INC
A Prentice Hall Company
Reston, Virginia

LIBRARY OF CONGRESS CATALOGING IN PUBLICATION DATA

Bugnolo, Dimitri Spartaco.
   Computer programs for electronic analysis and design.

   1. Electronic circuit-design--Computer programs.
2. Solid state electronics--Computer programs.
I. Title.
TK7867.B83 1983   621.3815'3'0285425   82-23167
ISBN 0-8359-0874-7

© 1983 by **Reston Publishing Company, Inc.**
         A Prentice-Hall Company
         Reston, Virginia 22090

10   9   8   7   6   5   4   3   2   1

Interior design and production: Jack Zibulsky

Printed in the United States of America

CONTENTS

Program

Program

PART II: PROBLEMS, 163–228

Twenty-seven Problems are included in order to illustrate the use of the programs in both analysis and design. They have been written so as to be of use to the individual user working alone, or in the classroom or micro-computer laboratory. Each problem provides an output listing from the particular program used in its solution, and extensive suggestions for the design of similar circuits. The problems should be of considerable value to the student.

PART III: APPENDICES

PREFACE

The world of solid-state electronics today encompasses everything from discrete transistors to very-large-scale integrated circuits. For the most part this is all based on silicon technology. Applications range from the simple one-transistor amplifier to the vast complexities of the large digital computer.

Texts in electronics intended for the undergraduate student or the hobbyist do not differ significantly from those of some twenty years ago. While the technology is up to date, the analysis is still based on the brute-force application of the engineering calculator to the equations; design is almost non-existent. There has been no large-scale effort to put the digital computer into the undergraduate texts in electronics.

It is my hope that this book at least partially solves this problem by creating a collection of computer programs that have been written in the BASIC Language so as to be immediately useful to the owner of any personal micro-computer (they all have BASIC capability); of course the programs will also run on the large-scale digital computers of the academic computing centers. I have tried to write a collection of programs for the analysis and, especially, the design of solid-state electronic circuits, that contain something of value for most levels of interest and expertise, from the beginning hobbyist or undergraduate student to the design engineer working in industry.

Programs 1 and 2 are intended for the beginner as an introduction to the simple fixed-bias common-emitter amplifier, and to the limitations of this circuit. Programs 3 and 4 design and analyze a common-emitter amplifier for its DC operating point. Program 4 should prove useful to the beginning student in electronics, because many different designs may be tested without the need for tedious calculations. This is a feature of all the design programs. The results of the design may be analyzed for the effects of the DC Beta spread in the population on the DC operating point.

Programs 5 through 15, 20 and 21 have been written so as to enable the user to actually design a solid-state single-stage amplifier to a given set of mid-band specifications, analyze the results, and then complete the design by choosing the coupling capacitors so as to meet a given set of specifications on the low-frequency 3dB point. Included are design programs for the common-emitter, the common-collector (emitter-follower), and the common-source amplifier. Programs to analyze the results of the user's designs are presented for the common-emitter, common-base, common-collector, common source, and common-gate amplifiers. These latter programs may also be used for design, though with less efficiency.

In addition to the programs covering the standard transistor amplifiers, there are a number of other programs which should be of interest to both the student and the practicing engineer. These include programs for solving a complex matrix equation by means of the modified Gauss-Jordan algorithm, a program for the design of a tuned-drain oscillator, a program to calculate the noise figure of a communications receiver, and a program to design a power supply based on a full-wave bridge rectifier.

Finally, for the user of integrated circuits, I have included a program to design low- and high-pass filters with integrated operational amplifiers, and a program to calculate the voltage-transfer function and drain current of an integrated C MOS inverter. The latter program is an excellent introduction to the power-saving abilities of the C MOS technology. These two programs should serve to introduce the student to integrated circuits, so that he or she may better be able to use large-scale computer programs, such as SPICE, that are intended for the analysis of small-, medium- and large-scale integrated circuits.

I have also included twenty-seven typical problems which may be solved with the programs in this book. Each solution contains the complete computer printouts, enabling the student to follow each step. The final problem is to design a complete three-stage amplifier.

The author welcomes correspondence from the reader suggesting new applications for these programs, or suggestions for programs which may be included in future editions of this book.

PART I
The Computer Programs

The programs in this book were written on the author's personal micro-computer system. This system consists of a Texas Instruments TI 99/4 computer and associated systems: a disk controller, a disk drive, and a thermal printer. This micro-computer uses a 16-bit micro-processor, running quadruple-precision, resulting in a 13-digit floating-point capability. Of this, 10 digits are actually printed. In view of this, the results shown are somewhat more accurate than those obtained on an 8-bit machine running double-precision or even a 16-bit machine running double-precision. I believe that the results for each program run, as printed in this book, should serve as benchmarks for applications of the programs.

PROGRAM **1**

Fixed-Bias
Common-Emitter
Transistor Amplifier;
the DC Operating Point

Analysis and simulation of a simple fixed-bias common-emitter bipolar-junction-transistor amplifier for its DC operating point. The program also allows the user to observe the effect that changing the base resistor has on the operating point. This program is designed for the beginner.

Circuit Description and Summary of the Analysis

The fixed-bias common-emitter BJT amplifier is a simple one-stage device which can be used as an introduction to transistor operation. The circuit is illustrated by Figure 1, for the case of an NPN transistor.

PROBLEM: _____

given V_{cc}, R_c, R_B, and the transistor's DC Beta;

calculate the transistor's operating point.

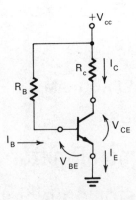

FIGURE 1. A fixed bias common emitter stage.

Applicable Equations:

$$I_B = (V_{cc} - V_{BE})/R_c, \qquad\qquad I_c = \beta_{DC} I_B$$

$$V_{BE} = V_{cc} - R_c I_c$$

Computer Procedure:

In order to illustrate the effect that changing the base resistor has on the operating point, the user should begin by inputing a large value for R_B, say 1 megohm. After inputing R_c, V_{cc} and the DC Beta, the user must specify the number of increments, N, between the chosen maximum value for the base resistor and the value, as calculated by the computer, of the resistor required to bring the transistor into saturation. For the examples that follow, N = 5 and N = 10 are used. The program is designed to yield a saturation value for V_{CE} of about 0.2 volts, a typical target for many transistors.

Enter Program 1.

Program Listing

```
10 PRINT "BJT CE FIXED BIAS-EFFE
CTS OF CHANGING RB ON OP POINT"
20 PRINT
30 INPUT "RB MAX= ":RB2
40 PRINT
50 INPUT "RC= ":RC
60 PRINT
```

```
70 INPUT "VCC= ":VCC
80 PRINT
90 INPUT "BETA DC= ":B
100 PRINT
110 INPUT "NO OF STEPS= ":N
120 PRINT
130 RB1=((VCC-.65)/(VCC-.2))*RC*
B
140 H=(RB2-RB1)/N
150 PRINT "      RB          IC
VCE"
160 PRINT
170 FOR M=0 TO N STEP 1
180 Y=RB2-H*M
190 IC=(B*(VCC-.65)/Y)
200 VCE=(VCC-(IC*RC))
210 PRINT INT(Y);TAB(10);(INT(IC
*1E6)/1E6);TAB(20);(INT(VCE*1E3)
/1E3)
220 NEXT M
230 END
```

Run

```
BJT CE FIXED BIAS-EFFECTS OF
CHANGING RB ON OP POINT

RB MAX=  1000000

RC=  2500

VCC=  25

BETA DC=  100

NO OF STEPS=  5
```

RB	IC	VCE
1000000	.002435	18.912
849092	.002867	17.83
698185	.003487	16.28
547278	.004449	13.876
396370	.006143	9.641
245463	.00992	.199

```
BJT CE FIXED BIAS-EFFECTS OF
CHANGING RB ON OP POINT

RB MAX=  1000000

RC=  2500

VCC=  25

BETA DC=  100

NO OF STEPS=  10

    RB        IC       VCE
 1000000   .002435   18.912
  924546   .002633   18.415
  849092   .002867   17.83
  773639   .003147   17.131
  698185   .003487   16.28
  622731   .00391    15.224
  547278   .004449   13.876
  471824   .00516    12.097
  396370   .006143   9.641
  320917   .007587   6.03
  245463   .00992    .199
```

Fixed-Bias
Common-Emitter
Transistor Amplifier;
Effects of Variations in DC
Beta on the Operating Point

Analysis and simulation of a simple fixed-bias common-emitter bipolar-junction-transistor amplifier for its DC operating point. The program allows the user to observe the effect that changing the DC Beta has on the operating point. This program is useful because the DC Beta of a transistor is usually specified by the manufacturer only to within a certain range of values: Beta Min ≤ DC Beta ≤ Beta Max.

Circuit Description and Summary of the Analysis

The user is referred to the corresponding section of Program 1.

PROBLEM: _____

    given V_{cc}, R_c, R_B, and the range of the transistor's DC Beta;

    calculate the transistor's operating point.

Computer Procedure:

This program may be used to illustrate the effects of various values of the transistor's DC Beta on the operating point of the circuit. The user must Input R_B, R_c, V_{cc}, Beta Max, and Beta Min. He must then specify the number of increments, N, he wishes to use to cover the range of the DC Beta. The program will print the DC operating point for N + 1 values of the DC Beta ranging from Beta Min to Beta Max.

Enter Program 2.

Program Listing

```
10 PRINT "BJT CE FIXED BIAS-EFFE
CTS OF CHANGE IN DC BETA ON OP"
20 PRINT
30 INPUT "RB= ":RB
40 PRINT
50 INPUT "RC= ":RC
60 PRINT
70 INPUT "VCC= ":VCC
80 PRINT
90 INPUT "BETA DC MAX= ":B2
100 PRINT
110 INPUT "BETA DC MIN= ":B1
120 PRINT
130 INPUT "NO OF STEPS ":N
140 PRINT
150 PRINT "BETA DC      IC
 VCE"
160 H=(B2-B1)/N
170 FOR M=0 TO N STEP 1
180 Y=B1+H*M
190 IC=(Y*(VCC-.65)/RB)
200 VCE=(VCC-(IC*RC))
210 PRINT
220 IF VCE<0 THEN 250
230 PRINT (INT(Y*1E3)/1E3);TAB(1
0);(INT(IC*1E6)/1E6);TAB(20);(IN
T(VCE*1E3)/1E3)
240 NEXT M
250 END
```

Run with following values:

$R_B = 1E6$

$R_c = 1500$

$$V_{cc} = 25$$

Beta Max = 500

Beta Min = 100

$$N = 10$$

Run

```
BJT CE FIXED BIAS-EFFECTS OF CHA
NGE IN DC BETA ON OP

BETA DC      IC        VCE
  100      .002435    21.347
  140      .003409    19.886
  180      .004383    18.425
  220      .005357    16.964
  260      .006331    15.503
  300      .007305    14.042
  340      .008279    12.581
  380      .009253    11.12
  420      .010227    9.659
  460      .011201    8.198
  500      .012175    6.737
```

PROGRAM **3**

Maximum and Minimum DC Operating Points for a Self-Biased Common-Emitter Transistor Amplifier

Analysis of a self-biased common-emitter bipolar-junction-transistor amplifier for the maximum and minimum values of its DC operating point, resulting from the specificied range of the transistor's DC Beta.

Circuit Description and Summary of the Analysis

The self-biased common-emitter BJT amplifier is the most often-used circuit in transistor systems. The circuit has the advantage of stabilizing the DC operating point despite population variations in the transistor's DC Beta. The circuit is illustrated by Figure 2 for the case of an NPN transistor.

PROBLEM: _____

> **given** V_{cc}, R_1, R_2, R_c, R_E, and Beta Min $\leq$ DC Beta $\leq$ Beta Max;
>
> **calculate** V-Thevenin, R-Thevenin, I_c Max, I_c Min, V_{CE} Max, V_{CE} Min.

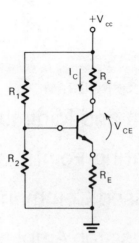

FIGURE 2. A self biased common emitter stage.

Note: Since the DC Beta is usually a function if I_c, this program may be repeated to obtain a better estimate of the operating point, given data for the dependence of Beta on I_c.

Applicable Equations:

$$V_{Th} = (V_{cc} * R_2) / (R_1 + R_2)$$

$$R_{Th} = R_1 \parallel R_2 = (R_1 * R_2) / (R_1 + R_2)$$

$$I_{c_{max}} = \beta_{DC_{max}} * (V_{Th} - V_o) / (R_{Th} + (\beta_{DC_{max}} + 1) * R_E)$$

$$I_{c_{min}} = \beta_{DC_{min}} * (V_{Th} - V_o) / (R_{Th} + (\beta_{DC_{min}} + 1) * R_E)$$

$$V_{CE_{max}} = V_{cc} - (R_c + R_E) * I_{c_{min}}$$

$$V_{CE_{min}} = V_{cc} - (R_c + R_E) * I_{c_{max}}$$

In the program, $V_o = V_{BE}$ is taken to be 0.65 volts. This is a good value for low- and medium-power transistors. Power transistors may be better modeled by using a value for V_{BE} of 0.70 volts. For this case the user may wish to change steps 95 and 100 in the program.

Note: This is a simplified program designed for beginners. For a more complete analysis the user should use Program 6.

Computer Procedure:

This program may be used to calculate the range of the DC operating point of a given self-biased common-emitter amplifier, given the range of the DC Beta specified by the manufacturer for the population. As such, the program is particularly useful for the analysis of production-line designs where the circuit must meet a given set of specifications, regardless of the value of the DC Beta within the specified range. The program may also be used to analyze a cirucit with a known DC Beta by setting both Beta Max and Beta Min equal to that single value.

Enter Program 3.

Program Listing

```
10 PRINT "MAX AND MIN OP POINT F
OR A  CE-BJT SELF BIASED CIRCUIT
"
15 PRINT
20 INPUT "R1= ":R1
25 PRINT
30 INPUT "R2= ":R2
35 PRINT
40 INPUT "RE= ":RE
45 PRINT
50 INPUT "RC= ":RC
55 PRINT
60 INPUT "VCC= ":VCC
65 PRINT
70 INPUT "BETA DC MAX= ":BMAX
75 PRINT
80 INPUT "BETA DC MIN= ":BMIN
85 VTH=(VCC*R2)/(R1+R2)
90 RTH=(R1*R2)/(R1+R2)
95 ICMAX=(BMAX*(VTH-.65))/(RTH+(
BMAX+1)*RE)
100 ICMIN=(BMIN*(VTH-.65))/(RTH+
(BMIN+1)*RE)
105 VCEMAX=(VCC-(RC+RE)*ICMIN)
110 VCEMIN=(VCC-(RC+RE)*ICMAX)
115 PRINT
120 PRINT "VTH= ";VTH
125 PRINT
130 PRINT "RTH= ";RTH
135 PRINT
140 PRINT "IC MAX= ";ICMAX
145 PRINT
150 PRINT "IC MIN= ";ICMIN
155 PRINT
160 PRINT "V-CE MAX= ";VCEMAX
165 PRINT
170 PRINT "V-CE MIN= ";VCEMIN
175 END
```

Run

```
MAX AND MIN OP POINT FOR A  CE-B
JT SELF BIASED CIRCUIT

R1=  39000
R2=  11000
RE=  5100
RC=  15000
VCC=  25
BETA DC MAX=  500
BETA DC MIN=  100

VTH=  5.5
RTH=  8580
IC MAX=  .0009459059
IC MIN=  .0009261381
V-CE MAX=  6.384624198
V-CE MIN=  5.987291706
```

Design of a Self-Biased Common-Emitter Transistor Amplifier with a Given DC Operating Point

A complete program for the DC design of a self-biased common-emitter transistor amplifier to achieve a specified upper and lower bound on the DC collector current. The design algorithm includes the effects that transistor population spread in the DC Beta and the reverse saturation current have on the operating point. After completing the initial design, the user may elect to specify 5%, 10%, or 20% resistors and calculate the acutal DC operating point for the nominal values of the chosen resistors.

Circuit Description and Summary of the Analysis

The design of a self-biased common-emitter BJT amplifier is a major element in the design of any solid-state system using BJT's. This circuit is the most commonly used, because of its operating-point stability despite differences in DC Beta among individual transistors from the same population.

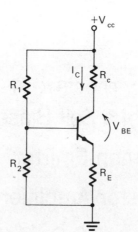

FIGURE 4. A self biased common emitter stage.

This is an advantage of great importance in designing for mass production, where the population statistics of any given transistor must be taken into consideration. The circuit is illustrated by Figure 4 for the case of an NPN transistor. The program may be used for PNP transistors by simply changing the signs of all PNP voltages, which are normally negative with respect to ground.

PROBLEM: _____

given V_{cc}, target V_{CE}, upper and lower bounds on I_c, DC Beta Max and Min, and I_{CEO};

calculate R_1, R_2, R_c, R_E, V_{CE} Max and Min, and I_c Max and Min.

Note: Since the DC Beta is usually a function of I_c, the user should use the transistor data for the average value of I_c.

Computer Procedure:

This is a design program with three parts. The first part will design a self-biased common-emitter transistor circuit to achieve certain collector characteristics: a given V_{CE}, and a given range for I_c. The design algorithm includes the effect that variation in the transistor's DC Beta has on the operating point, and the effect of the reverse saturation current. The computer will specify values for the circuit elements R_1, R_2, R_E, and R_c, and then calculate both the maximum and the minimum value of the collector-emitter voltage and the collector current.

At this point the user may elect to enter part two of the program. This part of the program is useful when the design is to be realized as part of a discrete-component system with 5%, 10%, or 20% resistors, or when the user wishes to specify a different set of nominal values for the resistors in an integrated-circuit realization of the computer design. The user must specify his choice of resistors based on the computer design. The program will then calculate the maximum and minimum values of the collector-emitter voltage and the collector current, based on the user's choice. If this result meets the original specifications, the user may elect to end the computer run at this point. If not, he may elect to repeat part two of this program, or return to the beginning and redo the initial design.

In effect, part three of this program is a repetition of part two with the additional information obtained from the initial effort. It may be used to calculate the effects of resistor population variations on the operating point target, by simply re-entering the values of any given resistor at its extremes.

The following two samples differ only in the value used for the reverse saturation current, I-CEO. In the first sample I-CEO = 5 μA, while in the second, I-CEO = 0. At or near room temperature, I-CEO may usually be neglected in high-Beta transistors. The 5 μA value used in sample number two is actually the maximum value expected.

In general, I-CEO is related to I-CBO, the reverse saturation current of the common-base configuration. The maximum value for I-CBO may often be found in transistor specification data. I-CEO may be computed from I-CBO by the equation:

$$I_{CEO} = (\beta_{DC} + 1) * I_{CBO}$$

For a worst-case test, the user should use I-CBO Max and β_{DC} Max.

Enter Program 4.

Program Listing

```
10 PRINT "PGM WILL DESIGN A CE-B
JT DC SELF-BIASED CIRCUIT FOR A
   GIVEN DESIGN TARGET."
20 PRINT
30 PRINT "USER MUST SPECIFY TARG
ET   V-CE AND RANGE OF I-C."
40 PRINT
50 INPUT "VCC?":VCC
60 PRINT
70 INPUT "TARGET V-CE?":VCE
```

Reference: Michael M. Cirovic, "Basic Electronics: Devices, Circuits and Systems", Second Edition, Reston Publishing Company, Inc., 1979. For a discussion of the design procedure used in this program, see Section 4.1.2.; the analysis part of the program uses conventional Hybrid-π theory, as given in Program 3.

```
80 PRINT
90 INPUT "I-C (UPPER,LOWER)?":IC
U,ICL
100 PRINT
110 INPUT "BETA DC (MAX,MIN)?":B
M,BN
120 PRINT
130 INPUT "I-CEO?":ICEO
140 VRE=VCE
150 IC=((ICU-ICL)/2+ICL)
160 RE=(VRE/IC)
170 RC=((VCC-2*VCE)/IC)
180 SM=(((ICU-ICL)*BM*BN)/(ICL*(
BM-BN)))
190 K=((SM*BM)/(BM+1-SM))
200 RTH=(K-1)*RE
210 IB=(ICL/BN)
220 VTH=((IB*(RTH+(BN+1)*RE))+.6
5)
230 R1=((RTH*VCC)/VTH)
240 R2=((R1*RTH)/(R1-RTH))
250 GOSUB 800
260 PRINT
270 PRINT "R1= ";R1
280 PRINT
290 PRINT "R2= ";R2
300 PRINT
310 PRINT "RE= ";RE
320 PRINT
330 PRINT "RC= ";RC
340 PRINT
350 PRINT "VCE MAX= ";VCEMAX
360 PRINT
370 PRINT "VCE MIN= ";VCEMIN
380 PRINT
390 PRINT "IC MAX= ";ICMAX
400 PRINT
410 PRINT "IC MIN= ";ICMIN
420 PRINT
430 PRINT "DO YOU WANT TO CHECK
YOUR DESIGN FOR ITS OPERATING RA
NGE WHEN YOU SPECIFY RESISTORS?"
440 PRINT
450 PRINT "IF YES ENTER 1, IF NO
 ENTER 0."
460 PRINT
470 INPUT A
480 IF A=1 THEN 500
490 IF A=0 THEN 790
500 PRINT
```

```
510 INPUT "R1?":R1
520 PRINT
530 INPUT "R2?":R2
540 PRINT
550 INPUT "RE?":RE
560 PRINT
570 INPUT "RC?":RC
580 VTH=(R2*VCC)/(R1+R2)
590 RTH=(R1*R2)/(R1+R2)
600 GOSUB 800
610 PRINT
620 PRINT "ICMAX=";ICMAX
630 PRINT
640 PRINT "ICMIN=";ICMIN
650 PRINT
660 PRINT "VCE MAX=";VCEMAX
670 PRINT
680 PRINT "VCE MIN=";VCEMIN
690 PRINT
700 PRINT "DOES YOUR 5,10,OR20%
RESISTOR DESIGN MEET SPECS ?"
710 PRINT
720 PRINT "IF NOT ENTER 1 TO RE-
SPECIFY RESISTORS"
730 PRINT
740 PRINT "IF SPECS. ARE MET THE
N ENTER 0 TO END PROGRAM."
750 PRINT
760 INPUT B
770 IF B=1 THEN 510
780 IF B=0 THEN 790
790 END
800 REM    SUBROUTINE #1
810 ICMAX=(((BM*(VTH-.65))+((RTH
+RE)*ICEO))/(RTH+(BM+1)*RE))
820 ICMIN=(((BN*(VTH-.65))+((RTH
+RE)*ICEO))/(RTH+(BN+1)*RE))
830 VCEMAX=((VCC-(RC*ICMIN))-((R
E*ICMIN*BN)/(BN+1)))
840 VCEMIN=((VCC-(RC*ICMAX))-((R
E*ICMAX*BM)/(BM+1)))
850 RETURN
```

Run Sample 1

```
PGM WILL DESIGN A CE-BJT DC SELF
-BIASED CIRCUIT FOR A   GIVEN DE
SIGN TARGET.

USER MUST SPECIFY TARGET    V-CE
```

```
AND RANGE OF I-C.

VCC=  25
TARGET V-CE=  5
UPPER BOUND IC=  .0011
LOWER BOUND IC=  .0009

BETA DC MAX=  500
BETA DC MIN=  100

I-CEO=  .000005

R1=  547650.7547

R2=  191248.8762

RE=  5000

RC=  15000

VCE MAX=  7.021920476

VCE MIN=  3.013416392

IC MAX=  .001099878

IC MIN=  .0009011345

DO YOU WANT TO CHECK YOUR DESIGN
 FOR ITS OPERATING RANGE WHEN YO
U SPECIFY RESISTORS?

IF YES ENTER 1, IF NO ENTER 0.

A=  1

R1=  560000

R2=  200000

RE=  5100

RC=  15000

ICMAX= .0010972324

ICMIN= .0008961289

VCE MAX= 7.033059743
```

VCE MIN= 2.956797339

DOES YOUR 5,10,OR20% RESISTOR DE
SIGN MEET SPECS ?

IF NOT ENTER 1 TO RE-SPECIFY RES
ISTORS

IF SPECS. ARE MET THEN ENTER 0 T
O END PROGRAM.

B=   0

Run Sample 2

PGM WILL DESIGN A CE-BJT DC SELF
-BIASED CIRCUIT FOR A   GIVEN DE
SIGN TARGET.

USER MUST SPECIFY TARGET   V-CE
 AND RANGE OF I-C.

VCC= 25
TARGET V-CE= 5
UPPER BOUND IC= .0011
LOWER BOUND IC= .0009

BETA DC MAX= 500
BETA DC MIN= 100

I-CEO= 0

R1= 547650.7547

R2= 191248.8762

RE= 5000

RC= 15000

VCE MAX= 7.044554455

VCE MIN= 3.018958092

IC MAX= .0010996008

IC MIN= .0009

DO YOU WANT TO CHECK YOUR DESIGN

FOR ITS OPERATING RANGE WHEN YO
U SPECIFY RESISTORS?

IF YES ENTER 1, IF NO ENTER 0.
A= 1

R1= 560000

R2= 200000

RE= 5100

RC= 15000

ICMAX= .0010969504

ICMIN= .0008949781

VCE MAX= 7.05613191

VCE MIN= 2.962464497

DOES YOUR 5,10,OR20% RESISTOR DE
SIGN MEET SPECS ?

IF NOT ENTER 1 TO RE-SPECIFY RES
ISTORS

IF SPECS. ARE MET THEN ENTER 0 T
O END PROGRAM.

B= 0

Design of a Self-Biased Common-Emitter Transistor Amplifier with a Given Minimum Mid-Band Overall Voltage Gain and a Minimum Input Impedance

A complete program for the simultaneous AC and DC design of a self-biased common-emitter amplifier for a specified minimum overall mid-band voltage gain and a minimum small-signal input impedance. The design algorithm includes the effect of the transistor junction temperature, and the effects of the DC or AC Beta on the gain and the input impedance. The program is written so as to interact with the designer, thus permitting the designer to specify such resistors as may be fixed by external considerations—for instance, the load seen by the amplifier. After completing the design, the program will print the design values for the circuit resistors and the minimum values of the overall mid-band voltage gain and input impedance.

Circuit Description and Summary of the Analysis

Like program 4, this program should be of particular value in the design of common-emitter amplifiers for mass production. In this case the program will design an amplifier to meet a system specification of the minimum overall small-signal voltage gain at mid-band, as well as a specification of the minimum value for the input impedance. In order to accomplish this, the algorithm was written so as to permit the designer to specify interactively other circuit parameters, such as the minimum DC and AC Beta's and the load resistor. Since the algorithm is written so as to minimize repetition it is best introduced by an actual run. The circuit is illustrated by Figure 5 for the case of an NPN transistor. The program may be used to design a PNP amplifier by simply changing the signs of all PNP voltages, which are normally negative with respect to ground.

PROBLEM: _____

    given Minimum specifications for the overall mid-band voltage gain
          and the input impedance (design criteria);

    calculate R_1, R_2, R_E, R_c, and the minimum values for the overall
          voltage gain and input impedance.

For the purpose of the program's algorithm, "mid-band" is defined in the usual manner, as that frequency range where the emitter is at AC ground and the reactances of C_1 and C_2 are such that they may be neglected in the calculations of the small-signal parameters.

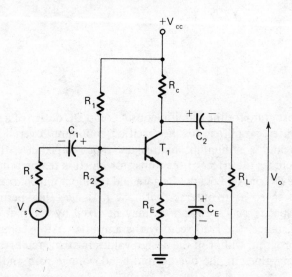

FIGURE 5. A single stage common emitter linear amplifier.

Computer Procedure:

The usefulness of this design program is best illustrated by RUNning the program. The PRINT statements will explicate the design algorithm. The user is required to make a number of decisions; they may or may not lead to a valid design. The program will signal the user if his choices lead to an invalid condition or result. This will become evident to the user as he RUNS the program. The example included in this program was chosen so as to illustrate what will occur should the user initially choose a collector resistor which is too small for the lower bound on the mid-band voltage gain. The initial choice was 6.8K ohms. Note that this resulted in a value of R-PI of less than 7.3K ohms, a result that cannot meet the specification on the minimum input impedance. The user was required to choose another value for R_c. It should be noted that the algorithm may be easily modified to override the specification for Z_{in} Min. by inserting a step such as: 465 GOTO 510.

The user is also required to specify the parameter GAMMA during the design. This parameter controls the values of resistors R_1 and R_2 by specifying the value of the current through resistor R_1 relative to the maximum value for the base current as predicted by the computer. Decreasing GAMMA will increase the values of both R_1 and R_2, thus leading to larger variations of the design results as a result of transistor population variations in the DC and AC Beta's. The user who is interested in the acutal range of the DC operating point and the mid-band voltage gain and input impedance should complete the computer design by choosing 5%, 10%, or 20% values for the resistors, or the nominal values for the integrated resistors, and proceed to Program 6.

The user should note that impossible design specifications may result in a negative value for $Z_{in \cdot min}$. Should this occur it is best to abort the program and begin again with a lower value for the overall voltage gain.

Applicable Equations:

$$A_v = \text{Overall voltage gain} = V_o / V_s$$
$$AV = g_m * (R_c \| R_L) * (Z_{in} / (Z_{in} + Z_s))$$
$$g_m = I_c / V_T \text{ where } V_T = (273 + T_J)/11600$$
$$I_{R_1} = \text{GAMMA} * I_{Bmax}; I_{Bmax} = I_c / \beta_{DCmin}$$
$$V_{RE} = \frac{1}{2} * (V_{cc} - I_c * R_c)$$
$$R_E = V_{RE} / I_c$$
$$R_1 = (V_{cc} - (V_{RE} + V_O))/I_{R_1}$$
$$R_2 = (V_{RE} + V_O)/(I_{R_1} - I_{Bmax})$$
$$R_{Th} = R_B = R_1 \| R_2$$
$$Z_{in} = r_\pi \| R_{Th}; r_\pi = \beta_{ACmin} / g_m$$

Note: The transistor junction temperature T_J must be in degrees centigrade.

Program 5 is based on an algorithm developed by the author and copyrighted in 1980.

Enter Program 5.

Program Listing

```
10 PRINT "COMMON EMMITER, MID-
BAND VOLTAGE AMPLIFIER DESIGN"
20 PRINT "ALL R IN OHMS"
30 PRINT "ALL I IN AMPS"
40 PRINT "ALL V IN VOLTS"
50 PRINT
60 PRINT "ENTER MINIMUM ALLOWABL
E VOLTAGE GAIN AT MIDBAND"
70 INPUT AV
80 PRINT
90 PRINT "ENTER MINIMUM ALLOWABL
E INPUT IMPEDANCE"
100 INPUT ZI
110 PRINT
120 PRINT "ENTER SOURCE IMPEDANC
E"
130 INPUT ZS
140 PRINT
150 PRINT "ENTER POWER SUPPLY VO
LTAGE"
160 INPUT VCC
170 PRINT
180 PRINT "ENTER JUNCTION TEMP.
IN DEG. C"
190 INPUT TJ
200 LWRZI=ZI/(ZI+ZS)
210 PRINT
220 PRINT "ENTER LOAD RESISTOR"
230 INPUT RL
240 PRINT
250 PRINT "ENTER COLLECTOR RESIS
TOR"
260 INPUT RC
270 RL1=RC*RL/(RC+RL)
280 GM=AV/(RL1*LWRZI)
290 PRINT
300 PRINT "THE LOWER BOUND ON GM
 IS GM=";GM
310 PRINT
320 VJ=(273+TJ)/11600
```

```
330 ICBL=GM*VJ
340 PRINT
350 PRINT "THE LOWER BOUND ON TH
E COLLECTOR CURRENT IS, IC=";ICB
L
360 PRINT
370 PRINT "CHOOSE A DESIGN VALUE
 FOR THE COLLECTOR CURRENT"
380 INPUT IC
390 PRINT
400 PRINT "ENTER MINIMUM TRANSIS
TOR SMALL SIGNAL BETA AT THE OPE
RATING POINT"
410 INPUT BAC
420 GM1=IC/VJ
430 RPI=BAC/GM1
440 PRINT
450 PRINT "THE SMALL SIGNAL TRAN
SISTOR INPUT-IMPEDANCE R-PIE=";
RPI
460 PRINT
470 IF (.9*RPI)<ZI THEN 480 ELSE
 520
480 PRINT "0.9*R-PIE<ZIN MIN. YO
UR DESIGN WILL PROBABLY FAIL. TR
Y INCREASING RC."
490 PRINT
500 GOTO 250
510 PRINT
520 VCE=.5*(VCC-(RC*IC))
530 IF VCE<.5 THEN 930
540 VRE=VCE
550 PRINT
560 PRINT "ENTER MINIMUM D.C.BET
A AT THE OPERATION POINT"
570 INPUT BDC
580 IBMAX=IC/BDC
590 PRINT
600 PRINT "DESIGNER MUST ENTER P
ARAMETER GAMMA"
610 INPUT GA
620 IR1=GA*IBMAX
630 R2=(VRE+0.65)/(IR1-IBMAX)
640 R1=(VCC-(VRE+.65))/IR1
650 RTH=R1*R2/(R1+R2)
660 ZIN=RPI*RTH/(RPI+RTH)
670 PRINT
680 PRINT "ZIN=";ZIN
690 PRINT
700 IF ZIN<ZI THEN 710 ELSE 730
```

```
710 PRINT "ZIN<SPEC. YOU LOSE. T
RY DECREASING GAMMA."
720 GOTO 590
730 RE=VRE/IC
740 AVS=((GM*RL1)*ZIN)/(ZIN+ZS)
750 AVI=(GM*RL1)
760 PRINT
770 PRINT "R1=";R1
780 PRINT
790 PRINT "R2=";R2
800 PRINT
810 PRINT "RE=";RE
820 PRINT
830 PRINT "RC=";RC
840 PRINT
850 PRINT "RL=";RL
860 PRINT
870 PRINT "AVI= ";AVI
880 PRINT "AVS=";AVS
890 PRINT
900 PRINT "THIS COMPLETES YOUR C
OMPUTER DESIGN. TO REPEAT FOR DI
FFERENT GAMMA, ENTER 1, TO END E
NTER 0."
910 INPUT X
920 IF X=1 THEN 580 ELSE 950
930 PRINT "VCE= ";VCE
940 PRINT "YOUR DESIGN SPECIFICA
TIONS ARE NOT REALISTIC"
950 END
```

Run

```
COMMON EMMITER, MID-  BAND VOLTA
GE AMPLIFIER DESIGN
ALL R IN OHMS
ALL I IN AMPS
ALL V IN VOLTS

ENTER MINIMUM ALLOWABLE VOLTAGE
GAIN AT MIDBAND
AV MIN=  200

ENTER MINIMUM ALLOWABLE INPUT IM
PEDANCE
ZIN MIN=  7000
```

```
ENTER SOURCE IMPEDANCE
ZS=  500

ENTER POWER SUPPLY VOLTAGE
VCC=  20

ENTER JUNCTION TEMP. IN DEG. C
TJ=  25

ENTER LOAD RESISTOR
RL=  7500

ENTER COLLECTOR RESISTOR
RC=  6800

THE LOWER BOUND ON GM IS GM=
 .0600840336

THE LOWER BOUND ON THE COLLECTOR
 CURRENT IS, IC= .0015435381

CHOOSE A DESIGN VALUE FOR THE CO
LLECTOR CURRENT
IC=  .0016

ENTER MINIMUM TRANSISTOR SMALL S
IGNAL BETA AT THE OPERATING POIN
T
MIN BETA AC=  450

THE SMALL SIGNAL TRANSISTOR INPU
T IMPEDANCE R-PIE=  7225.215517

0.9*R-PIE<ZIN MIN. YOUR DESIGN W
ILL PROBABLY FAIL. TRY INCREASIN
G RC.

ENTER COLLECTOR RESISTOR
RC=  8200

THE LOWER BOUND ON GM IS GM=
 .0547038328

THE LOWER BOUND ON THE COLLECTOR
 CURRENT IS, IC= .0014053226

CHOOSE A DESIGN VALUE FOR THE CO
LLECTOR CURRENT
IC=  .00145
```

```
ENTER MINIMUM TRANSISTOR SMALL S
IGNAL BETA AT THE OPERATING POIN
T
MIN BETA AC=  450

THE SMALL SIGNAL TRANSISTOR INPU
T IMPEDANCE R-PIE=  7972.651605


ENTER MINIMUM D.C.BETA AT THE OP
ERATION POINT
MIN BETA DC=  450

DESIGNER MUST ENTER PARAMETER GA
MMA
GAMMA=  40

ZIN= 6228.014199

ZIN<SPEC. YOU LOSE. TRY DECREASI
NG GAMMA.

DESIGNER MUST ENTER PARAMETER GA
MMA
GAMMA=  30

ZIN= 6595.888343

ZIN<SPEC. YOU LOSE. TRY DECREASI
NG GAMMA.

DESIGNER MUST ENTER PARAMETER GA
MMA
GAMMA=  20

ZIN= 7009.949569


R1= 237336.2069

R2= 76851.17967

RE= 2796.551724

RC= 8200

RL= 7500

AVI=  214.2857143
AVS= 200.0189265
```

```
THIS COMPLETES YOUR COMPUTER DES
IGN. TO REPEAT FOR DIFFERENT GAM
MA, ENTER 1, TO END ENTER 0.
X=  1

DESIGNER MUST ENTER PARAMETER GA
MMA
GAMMA=  15

ZIN= 7237.106632


R1= 316448.2759

R2= 104298.0296

RE= 2796.551724

RC= 8200

RL= 7500

AVI=  214.2857143
AVS= 200.4377913

THIS COMPLETES YOUR COMPUTER DES
IGN. TO REPEAT FOR DIFFERENT GAM
MA, ENTER 1, TO END ENTER 0.
X=  0
```

Analysis of a Self-Biased Common-Emitter Transistor Amplifier for the Maximum and Minimum Values of: Overall Mid-Band Voltage Gain, Input Impedance at Mid-Band, Collector-to-Emitter Voltage, and Collector Current

(This program may be used to analyze a design from Program 5.)

A complete program for the AC mid-band and DC analysis of a common-emitter self-biased transistor amplifier. This program may be used to find the operating point, given DC or AC Beta, or the range of operating points for a range of Beta's. Therefore, it may be used to analyze the results of the design of Program 5 for its maximum and minimum operating points as well as for the effects of specifying particular realizable resistors.

Circuit Description and Summary of the Analysis

The circuit for this program is again that of Figure 5. Since this is a straightforward program in analysis there is no special algorithm involved. Special features are the ability to specify the transistor's operating junction temperature. The program will automatically take into account the effect of junction temperature on the transconductance. However, the user should input a value for the reverse saturation current at the estimated junction temperature. The junction temperature of a transistor acting as a voltage amplifier, and drawing a few milliampers of current or less, may usually be taken as the circuit ambient temperature without causing excessive error.

PROBLEM: _____

given R_1, R_2, R_E, R_c, R_s, V_{cc}, the transistor's Max and Min DC and AC Beta's, junction temperature, and reverse saturation current;

calculate I_c Max and Min, V_{CE} Max and Min, Z_{in} Max and Min, and A_{vs} Max and Min.

Notes: Since the user can only INPUT one value for the reverse saturation current, it is recommended that the maximum possible value at the estimated junction temperature be used. The operating point resulting may then be compared to the operating point when I-CEO is taken equal to zero, in order to compute the effect of this usually small current.

Enter Program 6.

Program Listing

```
10 PRINT "ANALYSIS OF A COMMON E
MITTER BJT AMPLIFIER FOR RANGE O
F D.C. AND A.C. OP."
20 PRINT
30 INPUT "R1= ":R1
40 PRINT
50 INPUT "R2= ":R2
```

```
60 PRINT
70 INPUT "RE= ":RE
80 PRINT
90 INPUT "RC= ":RC
100 PRINT
110 INPUT "RL= ":RL
120 PRINT
130 INPUT "RS= ":RS
140 PRINT
150 INPUT "VCC= ":VCC
160 PRINT
170 INPUT "BETA DC (MAX,MIN)?":B
DM,BDN
180 PRINT
190 INPUT "BETA AC (MAX,MIN)?":B
AM,BAN
200 PRINT
210 INPUT "TEMP DEG. C= ":T
220 PRINT
230 INPUT "I-CEO= ":ICEO
240 PRINT
250 VTH=(R2*VCC)/(R1+R2)
260 RTH=(R1*R2)/(R1+R2)
270 ICMAX=(((BDM*(VTH-.65))+((RT
H+RE)*ICEO))/(RTH+(BDM+1)*RE))
280 ICMIN=(((BDN*(VTH-.65))+((RT
H+RE)*ICEO))/(RTH+(BDN+1)*RE))
290 VCEMAX=((VCC-(RC*ICMIN))-((R
E*ICMIN*BDN)/(BDN+1)))
300 VCEMIN=((VCC-(RC*ICMAX))-((R
E*ICMAX*BDM)/(BDM+1)))
310 PRINT "ICMAX= ";ICMAX
320 PRINT "ICMIN= ";ICMIN
330 PRINT
340 PRINT "VCE MAX= ";VCEMAX
350 PRINT "VCE MIN= ";VCEMIN
360 PRINT
370 VT=((T+273)/11600)
380 GMM=(ICMAX/VT)
390 GMN=(ICMIN/VT)
400 RPM=(BAM/GMM)
410 RPN=(BAN/GMN)
420 RL1=(RC*RL/(RC+RL))
430 ZINM=(RPM*RTH/(RPM+RTH))
440 ZINN=(RPN*RTH/(RPN+RTH))
450 AVIM=(GMM*RL1)
460 AVIN=(GMN*RL1)
470 AVSM=((GMM*RL1)*ZINM)/(ZINM+
RS)
480 AVSN=((GMN*RL1)*ZINN)/(ZINN+
RS)
490 PRINT "ZIN MAX= ";ZINM
```

```
500 PRINT "ZIN MIN= ";ZINN
510 PRINT
520 PRINT "AVI MAX= ";AVIM
530 PRINT "AVI MIN= ";AVIN
540 PRINT
550 PRINT "AVS MAX= ";AVSM
560 PRINT "AVS MIN= ";AVSN
570 END
```

The following is an analysis of the design of example of Program 5.

Run

```
ANALYSIS OF A COMMON EMITTER BJT
 AMPLIFIER FOR RANGE OF D.C. AND
 A.C. OP.

R1=  240000

R2=  82000

RE=  3000

RC=  8200

RL=  7500

RS=  500

VCC=  20

BETA DC MAX=  1200
BETA DC MIN=  450

BETA AC MAX=  1400
BETA AC MIN=  450

TEMP DEG. C=  25

I-CEO=  0

ICMAX=  .0014551391
ICMIN=  .0014139028

VCE MAX=  4.173693374
VCE MIN=  3.706077341

ZIN MAX=  17599.10473
ZIN MIN=  7211.465269

AVI MAX=  221.8818035
AVI MIN=  215.5940414

AVS MAX=  215.7521687
AVS MIN=  201.6152427
```

Analysis of a Self-Biased Common-Base Transistor Amplifier for the Maximum and Minimum Values of: Overall Mid-Band Voltage Gain, Input Impedance at Mid-Band, Current Gain, Power Gain, Collector-to-Emitter Voltage, and Collector Current

A complete program for the AC mid-band and DC analysis of a common-base self-biased transistor amplifier. This program may be used to find the operating point for a given DC or AC Beta, or the range of operating points for a range of Beta's. The program will also accept an estimate for the junction temperature and a value for the reverse saturation current.

Circuit Description and Summary of the Analysis

The common-base amplifier has the advantage of having a small controllable input impedance and finds applications where the common-emitter circuit cannot be used. The circuit for this program is illustrated by Figure 7. This program may be used for designing an amplifier with a particular input impedance, by changing the circuit parameters in a logical fashion. The input impedance is given by:

$$\frac{1}{Z_{in}} = \left[\frac{1}{R_E} + \frac{1}{r_\pi} + g_m \right].$$

PROBLEM: _____

given R_1, R_2, R_E, R_c, R_s, V_{cc}, and the transistor's Max and Min DC and AC Beta's, junction temperature, and reverse saturation current;

calculate I_c Max and Min, V_{CE} Max and Min, Z_{in} Max and Min, current gain Max and Min, power gain Max and Min and the Max and Min of the mid-band overall voltage gain, A_{Vs}, ($= V_o/V_s$).

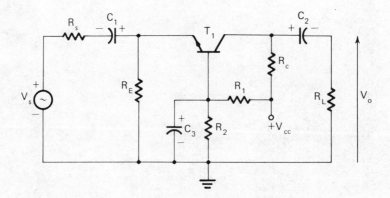

FIGURE 7. A single stage common base linear amplifier.

Enter Program 7.

Program Listing

```
10 PRINT "ANALYSIS OF A COMMON B
ASE BJT AMPLIFIER FOR RANGE OF D
.C. AND A.C. OP."
20 PRINT
30 INPUT "R1= ":R1
40 PRINT
50 INPUT "R2= ":R2
60 PRINT
70 INPUT "RE= ":RE
80 PRINT
90 INPUT "RC= ":RC
100 PRINT
110 INPUT "RL= ":RL
120 PRINT
130 INPUT "RS= ":RS
140 PRINT
150 INPUT "VCC= ":VCC
160 PRINT
170 INPUT "BETA DC (MAX,MIN)?":B
DM,BDN
180 PRINT
190 INPUT "BETA AC (MAX,MIN)?":B
AM,BAN
200 PRINT
210 INPUT "TEMP DEG. C= ":T
220 PRINT
230 INPUT "I-CEO= ":ICEO
240 PRINT
250 VTH=(R2*VCC)/(R1+R2)
260 RTH=(R1*R2)/(R1+R2)
270 ICMAX=(((BDM*(VTH-.65))+((RT
H+RE)*ICEO))/(RTH+(BDM+1)*RE))
280 ICMIN=(((BDN*(VTH-.65))+((RT
H+RE)*ICEO))/(RTH+(BDN+1)*RE))
290 VCEMAX=((VCC-(RC*ICMIN))-((R
E*ICMIN*BDN)/(BDN+1)))
300 VCEMIN=((VCC-(RC*ICMAX))-((R
E*ICMAX*BDM)/(BDM+1)))
310 PRINT "ICMAX= ";ICMAX
320 PRINT "ICMIN= ";ICMIN
330 PRINT
340 PRINT "VCE MAX= ";VCEMAX
350 PRINT "VCE MIN= ";VCEMIN
360 PRINT
370 VT=((T+273)/11600)
```

```
380 GMM=(ICMAX/VT)
390 GMN=(ICMIN/VT)
400 RPM=(BAM/GMM)
410 RPN=(BAN/GMN)
420 RL1=(RC*RL/(RC+RL))
430 YINM=((1/RE)+(1/RPM)+GMM)
440 ZINM=1/YINM
450 YINN=((1/RE)+(1/RPN)+GMN)
460 ZINN=1/YINN
470 AVIM=GMM*RL1
480 AVIN=GMN*RL1
490 AVSM=AVIM*ZINM/(ZINM+RS)
500 AVSN=AVIN*ZINN/(ZINN+RS)
510 AIM=AVIM*ZINM/RL
520 AIN=AVIN*ZINN/RL
530 APM=AVIM*AIM
540 APN=AVIN*AIN
550 PRINT "ZIN MAX= ";ZINM
560 PRINT "ZIN MIN= ";ZINN
570 PRINT
580 PRINT "AVI MAX= ";AVIM
590 PRINT "AVI MIN= ";AVIN
600 PRINT
610 PRINT "AVS MAX= ";AVSM
620 PRINT "AVS MIN= ";AVSN
630 PRINT
640 PRINT "MAX CURRENT GAIN= ";A
IM
650 PRINT "MIN CURRENT GAIN= ";A
IN
660 PRINT
670 PRINT "MAX POWER GAIN= ";APM
680 PRINT "MIN POWER GAIN= ";APN
690 PRINT
700 PRINT "OUTPUT Z= ";RC
710 END
```

Run

```
ANALYSIS OF A COMMON BASE BJT AM
PLIFIER FOR RANGE OF D.C. AND A.
C. OP.

R1=  51000

R2=  24000

RE=  5100

RC=  5100
```

```
RL=  2000

RS=  50

VCC=  15

BETA DC MAX=  1200
BETA DC MIN=  450

BETA AC MAX=  1400
BETA AC MIN=  450

TEMP DEG. C=  25

I-CED=  0

ICMAX=  .0008108874
ICMIN=  .0008062009

VCE MAX=  6.785867015
VCE MIN=  6.732392082

ZIN MAX=  31.46299613
ZIN MIN=  31.59743782

AVI MAX=  45.34653338
AVI MIN=  45.08445797

AVS MAX=  17.51393727
AVS MIN=  17.45830991

MAX CURRENT GAIN=  .713368902
MIN CURRENT GAIN=  .7122766787

MAX POWER GAIN=  32.34880673
MIN POWER GAIN=  32.11260798

OUTPUT Z=  5100
```

Analysis of Self-Biased Emitter-Follower Transistor Amplifier for the Maximum and Minimum Values of: Overall Mid-Band Voltage Gain, Input and Output Impedances at Mid-Band, Current Gain, Power Gain, and Junction Temperature

A complete program for the AC mid-band and DC analysis of a self-biased emitter-follower transistor amplifier. This program may be used to find the operating point for a given DC or AC Beta, or the range of operating points for a range of Beta's. The program will also calculate the maximum and minimum junction temperature of the transistor given the junction-ambient thermal resistance.

Circuit Description and Summary of the Analysis

The emitter-follower amplifier has the advantage of having a small controllable output impedance, and finds applications where the common-emitter or common-base amplifier cannot be used. The circuit for this program is illustrated by Figure 8.

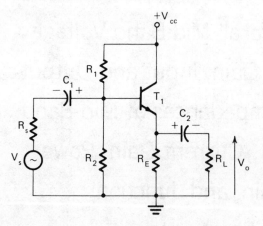

FIGURE 8. A single stage emitter follower linear amplifier.

This program may be used for designing an amplifier with a particular output impedance, by changing the circuit parameters in a logical fashion. The output impedance is given by:

$$Z_o = R_E \parallel \left[\frac{r_\pi + R_{Th} \parallel R_s}{1 + \beta_{AC}} \right]$$

PROBLEM: _____

> **given** R_1, R_2, R_E, R_s, V_{cc}, T(ambient), thermal resistance (junction to ambient), reverse saturation current, and the maximum and minimum values of the DC and AC Beta's;
>
> **calculate** the maximum and minimum values of: I_c, V_{CE}, Z_{in}, T_J, A_{Vi}, A_{Vs}, A_I, A_P, and Z_o.

The following is an example of an emitter-follower circuit that has been designed to have an output impedance of about 52 ohms. The transistor specifications used are similar to those of the 2N5089.

Enter Program 8.

Program Listing

```
10 PRINT "ANALYSIS OF AN EMITTER
 FOLLOWER BJT AMPLIFIER FOR RANG
E OF D.C. AND A.C. OP"

20 PRINT
30 INPUT "R1= ":R1
40 PRINT
50 INPUT "R2= ":R2
60 PRINT
70 INPUT "RE= ":RE
80 PRINT
90 INPUT "RL= ":RL
100 PRINT
110 INPUT "RS= ":RS
120 PRINT
130 INPUT "VCC= ":VCC
140 PRINT
150 INPUT "BETA DC (MAX,MIN)?":B
DM,BDN
160 PRINT
170 INPUT "BETA AC (MAX,MIN)?":B
AM,BAN
180 PRINT
190 INPUT "AMBIENT TEMP DEG C= "
:TA
200 PRINT
210 INPUT "THERMAL RESISTANCE. J
UNCTION TO AMBIANT, IN DEG C/WAT
T =":TR
220 PRINT
230 INPUT "I-CEO= ":ICEO
240 PRINT
250 VTH=(R2*VCC)/(R1+R2)
260 RTH=(R1*R2)/(R1+R2)
270 ICMAX=(((BDM*(VTH-.65))+((RT
H+RE)*ICEO))/(RTH+(BDM+1)*RE))
280 ICMIN=(((BDN*(VTH-.65))+((RT
H+RE)*ICEO))/(RTH+(BDN+1)*RE))
290 VCEMAX=((VCC-(RC*ICMIN))-((R
E*ICMIN*BDN)/(BDN+1)))
300 VCEMIN=((VCC-(RC*ICMAX))-((R
E*ICMAX*BDM)/(BDM+1)))
```

```
310 PRINT "ICMAX= ";ICMAX
320 PRINT "ICMIN= ";ICMIN
330 PRINT
340 PRINT "VCE MAX= ";VCEMAX
350 PRINT "VCE MIN= ";VCEMIN
360 PRINT
370 TJM=(TA+TR*VCEMIN*ICMAX)
380 TJN=(TA+TR*VCEMAX*ICMIN)
390 VTM=(TJM+273)/11600
400 VTN=(TJN+273)/11600
410 GMM=(ICMAX/VTM)
420 GMN=(ICMIN/VTN)
430 RPM=(BAM/GMM)
440 RPN=(BAN/GMN)
450 RL1=(RE*RL/(RE+RL))
460 Z1M=(RPM+(1+BAM)*RL1)
470 Z1N=(RPN+(1+BAN)*RL1)
480 ZINM=(RTH*Z1M)/(RTH+Z1M)
490 ZINN=(RTH*Z1N)/(RTH+Z1N)
500 AVIM=((1+BAM)*RL1)/Z1M
510 AVIN=((1+BAN)*RL1)/Z1N
520 AVSM=(AVIM*ZINM)/(ZINM+RS)
530 AVSN=(AVIN*ZINN)/(ZINN+RS)
540 AIM=(AVIM*ZINM)/RL
550 AIN=(AVIN*ZINN)/RL
560 APM=AVIM*AIM
570 APN=AVIN*AIN
580 RSB=(RS*RTH)/(RS+RTH)
590 ZO1M=(RPM+RSB)/(1+BAM)
600 ZO1N=(RPN+RSB)/(1+BAN)
610 ZOM=(RE*ZO1M)/(RE+ZO1M)
620 ZON=(RE*ZO1N)/(RE+ZO1N)
630 PRINT "T-J MAX DEG C =";TJM
640 PRINT "T-J MIN DEG C =";TJN
650 PRINT
660 PRINT "ZIN MAX= ";ZINM
670 PRINT "ZIN MIN= ";ZINN
680 PRINT
690 PRINT "AVI MAX= ";AVIM
700 PRINT "AVI MIN= ";AVIN
710 PRINT
720 PRINT "AVS MAX= ";AVSM
730 PRINT "AVS MIN= ";AVSN
740 PRINT
750 PRINT "MAX CURRENT GAIN= ";A
IM
760 PRINT "MIN CURRENT GAIN= ";A
IN
770 PRINT
780 PRINT "MAX POWER GAIN= ";APM
```

```
790 PRINT "MIN POWER GAIN= ";APN
800 PRINT
810 PRINT "MAX OUTPUT Z= ";ZON
820 PRINT "MIN OUTPUT Z= ";ZOM
830 END
```

Run

```
ANALYSIS OF AN EMITTER FOLLOWER
BJT AMPLIFIER FOR RANGE OF D.C.
AND A.C. OP

R1=  56000

R2=  27000

RE=  8200

RL=  52

RS=  1000

VCC=  15

BETA DC MAX=  1200
BETA DC MIN=  450

BETA AC MAX=  1400
BETA AC MIN=  450

AMBIENT TEMP IN DEG C =  25

THERMAL RESISTANCE, JUNCTION TO
AMBIENT, IN DEG. C/WATT=  360

I-CEO=  0

ICMAX=  .0005144139
ICMIN=  .0005121285

VCE MAX=  10.80985743
VCE MIN=  10.7853185

T-J MAX DEG C = 26.99732228
T-J MIN DEG C = 26.99297313

ZIN MAX=  16155.58306
ZIN MIN=  13051.43322

AVI MAX=  .5070347146
AVI MIN=  .5063017773
```

```
AVS MAX=  .4774796296
AVS MIN=  .4702697392

MAX CURRENT GAIN=  157.5277201
MIN CURRENT GAIN=  127.0762276

MAX POWER GAIN=  79.87202262
MIN POWER GAIN=  64.3389199

MAX OUTPUT Z=  52.15409937
MIN OUTPUT Z=  50.60093811
```

Complete Analysis of the Common-Emitter Self-Biased Transistor Amplifier for its: DC Operating Point; Mid-Band Voltage, Current, and Power Gains; Mid-Band Input Impedance; Low-Frequency and High-Frequency 3dB Points; and Junction Temperature

(The program will also calculate the maximum and minimum of the above, given the bounds on the DC and AC Betas.)

Circuit Description, Summary of the Analysis, and Discussion of the Program

The common-emitter self-biased transistor amplifier is the most often-used solid-state design. It is characterized by moderately high values of voltage gain and input impedance, a very large power gain, a moderate value of current gain, and an output impedance equal to the collector resistor. The circuit used for this program is shown in Figure 5 (see Program 5).

Program 9 may be used in conjunction with Program 5 to completely design a common-emitter amplifier. The very experienced designer may find the inputs obtainable from Program 5 unnecessary. However, it should be noted that the algorithm used in Program 5 is difficult to duplicate in an analysis program without a great deal of very practical, hands-on experience in the laboratory. The user is urged to use Program 5 initially, and then move on to Program 9 for "fine-tuning" and for low- and high-frequency analysis or design.

The program is constructed so as to permit the user to repeat the low- and high-frequency analysis without reINPUTing all the other parameters. This construction simplifies the use of the program for design.

The effects of stray capacitance to ground or to the sub-strata are not included directly in this program, as these will depend on individual physical realizations of the results of the design and analysis. However, if required at this stage, any stray capacitance from the base to "ground" may be included in the INPUT for C-PI. Any stray from the collector to "ground" may be calculated by using the equation:

$$F_s = 1/(2\pi \, RL_1 \cdot C_s),$$

where RL_1 may be found in Step 600 of the program.

The program also takes into account the effect of the transistor's operating point on the junction temperature. Both the maximum and minimum values of the junction temperature are printed. The user must INPUT the ambient temperature in degrees Centigrade and the junction-to-ambient thermal resistance. The effects of a heat sink can be modeled at this point.

PROBLEM: _____

> **given** R_1, R_2, R_E, R_c, R_L, and R_s, all in ohms; C_1, C_2, C_E, C-Pi, and C-Mu, all in farads; V_{cc} in volts; DC Beta and AC Beta, Max and Min; ambient temperature in degrees Centigrade; thermal Resistance (junction to ambient) in degrees Centigrade per watt; and reverse saturation current in amperes;
>
> **calculate** the complete design of the circuit.

Notes: Engineering notation should be used to facilitate the entry of the above INPUTS. This program may be used for a single Beta by taking Beta Max equal to Beta Min. The junction-temperature prediction and the reverse-saturation-current effects may be omitted by simply entering a zero at the appropriate points in the INPUTs.

Sample Program:

The sample program was chosen to be similar to the circuit used in Program 6. This feature permits the reader to note the effect that the operating point of the circuit has on the junction temperature, and the effect that the very small (in this case, because V_{CE} is kept small to achieve a very large gain) change in the junction temperature has on the various gains and 3dB cut-off frequencies. These are PRINTed in Hz rather than in the radians/sec. found in most references.

Enter Program 9.

Program Listing

```
10 PRINT "COMPLETE ANALYSIS OF A
   COMMON EMITTER AMPLIFIER"
15 PRINT
20 INPUT "R1= ":R1
25 PRINT
30 INPUT "R2= ":R2
35 PRINT
40 INPUT "RE= ":RE
45 PRINT
50 INPUT "RC= ":RC
55 PRINT
60 INPUT "RL= ":RL
65 PRINT
70 INPUT "RS= ":RS
75 PRINT
80 INPUT "C1= ":C1
85 PRINT
90 INPUT "C2= ":C2
95 PRINT
100 INPUT "CE= ":CE
105 IF K=1 THEN 355
110 PRINT
115 INPUT "C PIE= ":CP
120 PRINT
```

```
125 INPUT "C MU= ":CU
130 IF K=2 THEN 395
135 PRINT
140 INPUT "VCC= ":VCC
145 PRINT
150 INPUT "BETA DC (MAX,MIN)?":B
DM,BDN
155 PRINT
160 INPUT "BETA AC (MAX,MIN)?":B
AM,BAN
165 PRINT
170 INPUT "AMBIENT TEMP DEG C= "
:TA
175 PRINT
180 INPUT "THERMAL RESISTANCE. J
UNCTION TO AMBIENT, IN DEG C/WAT
T =":TR
185 PRINT
190 INPUT "I-CEO= ":ICEO
195 PRINT
200 VTH=(R2*VCC)/(R1+R2)
205 RTH=(R1*R2)/(R1+R2)
210 ICMAX=(((BDM*(VTH-.65))+((RT
H+RE)*ICEO))/(RTH+(BDM+1)*RE))
215 ICMIN=(((BDN*(VTH-.65))+((RT
H+RE)*ICEO))/(RTH+(BDN+1)*RE))
220 VCEMAX=((VCC-(RC*ICMIN))-((R
E*ICMIN*BDN)/(BDN+1)))
225 VCEMIN=((VCC-(RC*ICMAX))-((R
E*ICMAX*BDM)/(BDM+1)))

230 PRINT· "ICMAX= ";ICMAX
235 PRINT "ICMIN= ";ICMIN
240 PRINT
245 PRINT "VCE MAX= ";VCEMAX
250 PRINT "VCE MIN= ";VCEMIN
255 PRINT
260 TJN=(TA+TR*VCEMIN*ICMAX)
265 TJM=(TA+TR*VCEMAX*ICMIN)
270 VTM=(TJM+273)/11600
275 VTN=(TJN+273)/11600
280 GMM=(ICMAX/VTM)
285 GMN=(ICMIN/VTN)
290 RPM=(BAM/GMM)
295 RPN=(BAN/GMN)
300 RL1=(RC*RL/(RC+RL))
305 ZINM=(RPM*RTH/(RPM+RTH))
310 ZINN=(RPN*RTH/(RPN+RTH))
315 AVIM=(GMM*RL1)
320 AVIN=(GMN*RL1)
325 AVSM=(AVIM*ZINM)/(ZINM+RS)
```

```
330 AVSN=(AVIN*ZINN)/(ZINN+RS)
335 AIM=(AVIM*ZINM)/RL
340 AIN=(AVIN*ZINN)/RL
345 APM=(AIM*AVIM)
350 APN=(AIN*AVIN)
355 FC1M=((1/((RS+ZINM)*C1))/6.2
83185)
360 FC1N=((1/((RS+ZINN)*C1))/6.2
83185)
365 FC2=((1/((RC+RL)*C2))/6.2831
85)
370 RSB=(RS*RTH/(RS+RTH))
375 Z1M=((RPM+RSB)/(1+BAM))
380 Z1N=((RPN+RSB)/(1+BAN))
385 RTM=(RE*Z1M/(RE+Z1M))
390 RTN=(RE*Z1N/(RE+Z1N))
395 FCEM=((1/(RTM*CE))/6.283185)
400 FCEN=((1/(RTN*CE))/6.283185)
405 IF K=1 THEN 550
410 RSM=(ZINM*RS/(ZINM+RS))
415 RSN=(ZINN*RS/(ZINN+RS))
420 CINM=(CP+(CU*((1+GMM*RL1+(RL
1/RSM))))
425 CINN=(CP+CU*((1+GMN*RL1+(RL1
/RSN))))
430 FPM=((1/(RSN*CINN))/6.283185
)
435 FPN=((1/(RSM*CINM))/6.283185
)
440 IF K=2 THEN 590
445 PRINT "T-J MAX DEG C =";TJM
450 PRINT "T-J MIN DEG C =";TJN
455 INPUT "ENTER 1 TO PROCEED":G
460 PRINT
465 PRINT "ZIN MAX= ";ZINM
470 PRINT "ZIN MIN= ";ZINN
475 PRINT
480 PRINT "AVS MAX= ";AVSM
485 PRINT "AVS MIN= ";AVSN
490 PRINT
495 PRINT "AVI MAX= ";AVIM
500 PRINT "AVI MIN= ";AVIN
505 PRINT
510 PRINT "MAX CURRENT GAIN= ";A
IM
515 PRINT "MIN CURRENT GAIN= ";A
IN
520 PRINT
525 PRINT "MAX POWER GAIN= ";APM
530 PRINT "MIN POWER GAIN= ";APN
```

```
535 INPUT "ENTER 1 TO PROCEED":H
540 PRINT
545 PRINT "OUTPUT Z= ";RL
550 PRINT
555 PRINT "F C1 MAX= ";FC1N
560 PRINT "F C1 MIN= ";FC1M
565 PRINT
570 PRINT "F C2= ";FC2
575 PRINT
580 PRINT "F CE MAX= ";FCEM
585 PRINT "F CE MIN= ";FCEN
590 PRINT
595 IF K=1 THEN 610
600 PRINT "MAX H.F. 3 D.B. POINT
 = ";FPM
605 PRINT "MIN H.F. 3 D.B. POINT
 = ";FPN
610 PRINT
615 PRINT "ENTER 1 TO REPEAT LOW
 FREQ ANALYSIS"
620 PRINT "ENTER 2 TO REPEAT HIG
H FREQ ANALYSIS"
625 PRINT "ENTER 0 TO END"
630 INPUT K
635 IF K=0 THEN 660
640 IF K=1 THEN 75
645 IF K=2 THEN 110
650 PRINT "TRY AGAIN"
655 GOTO 610
660 END
```

Run

```
COMPLETE ANALYSIS OF A COMMON EM
ITTER AMPLIFIER

R1=  270000

R2=  62000

RE=  2000

RC=  10000

RL=  6800

RS=  500

C1=  .000006
```

```
C2=  .00001

CE=  .0001

C PIE=  1.5E-11

C MU=  2.E-12

VCC=  20
BETA DC MAX=  1200
BETA DC MIN=  450

BETA AC MAX=  1400
BETA AC MIN=  450

AMBIENT TEMP. IN DEG C = 25

THERMAL RESISTANCE, JUNCTION TO
AMBIENT, IN DEG. C/WATT =  360

I-CEO=  0


ICMAX=  .0015094988
ICMIN=  .0014575717

VCE MAX=  2.515603402
VCE MIN=  1.888527561

T-J MAX DEG C = 26.32000203
T-J MIN DEG C = 26.02626286

ZIN MAX=  16228.92644
ZIN MIN=  6873.618946

AVS MAX=  229.7081409
AVS MIN=  213.3452484

AVI MAX=  236.7852615
AVI MIN=  228.864384


MAX CURRENT GAIN=  565.1133224
MIN CURRENT GAIN=  231.3421421

MAX POWER GAIN=  133810.5058
MIN POWER GAIN=  52945.97684

OUTPUT Z=  10000
```

```
F C1 MAX=  3.597395708
F C1 MIN=  1.585626264

F C2=  .9473508981

F CE MAX=  92.07930331
F CE MIN=  87.09071205

MAX H.F.  3 D.B.  POINT=
 693896.4979
MIN H.F.  3 D.B.  POINT=
 646841.6402
```

A Simplified Low- and High-Frequency Analysis for the Common-Emitter Transistor Amplifier

A simplified analysis for the low- and high-frequency 3dB points of a common-emitter amplifier that may be easily used when the accuracy of Program 9 is not required. This program neglects such design factors as Beta spread; it also assumes that the circuit is operating at a very low power level, so that the junction temperature may be assumed to be equal to room temperature (25 degrees Centigrade). The collector current must be INPUTed. It may be obtained from Program 6 or any other convenient source. This program is short and easy to run, but never the less saves a great deal of time on the engineering calculator. The circuit is illustrated by Figure 5 (see Program 5).

Sample Program:

The sample used to illustrate this program is similar to that already used for Programs 6 and 9. This leads to consistency in the presentation and an understanding of the differences.

For example, we note that the low-frequency results of Program 9 differ from those of this program by only about $\pm 3\%$, and the high-frequency results of Program 9 differ from those of this program by only about $\pm 5\%$. These rather small differences between the results of the complete Program 9 and this simple Program 10 are deceptive, and due to the fact that the author has chosen a circuit of high stability against Beta variations. When in doubt, it is best to use Program 9.

Enter Program 10.

Program Listing

```
10 PRINT "A SIMPLIFIED LOW AND H
IGH FREQ. ANALYSIS FOR THE COMMO
N EMITTER AMPLIFIER."
15 PRINT
20 INPUT "R1= ":R1
25 PRINT
30 INPUT "R2= ":R2
35 PRINT
40 INPUT "RE= ":RE
45 PRINT
50 INPUT "RC= ":RC
55 PRINT
60 INPUT "RL= ":RL
65 PRINT
70 INPUT "RS= ":RS
75 PRINT
80 INPUT "C1= ":C1
85 PRINT
90 INPUT "C2= ":C2
95 PRINT
100 INPUT "CE= ":CE
105 PRINT
110 INPUT "C PIE= ":CP
115 PRINT
120 INPUT "C MU= ":CU
125 PRINT
130 INPUT "I-C D.C.= ":IC
135 PRINT
140 INPUT "BETA A.C.= ":B
145 RTH=(R1*R2)/(R1+R2)
150 GM=(IC/.02569)
155 RP=(B/GM)
160 RL1=(RC*RL)/(RC+RL)
165 ZIN=(RP*RTH)/(RP+RTH)
170 FC1=((1/((RS+ZIN)*C1))/6.283
185)
175 FC2=((1/((RC+RL)*C2))/6.2831
```

```
85)
180 RSB=(RS*RTH)/(RS+RTH)
185 Z1=((RP+RSB)/(1+B))
190 RT=(RE*Z1)/(RE+Z1)
195 FCE=((1/(RT*CE))/6.283185)
200 RS1=(ZIN*RS)/(ZIN+RS)
205 CIN=(CP+CU*((1+GM*RL1+(RL1/R
S1))))
210 FP=((1/(RS1*CIN))/6.283185)
215 PRINT
220 PRINT "THE LOW FREQUENCY E D
B POINT IS DETERMINED BY THE LAR
GER OF THE FOLLOWING THREE FREQU
ENCIES IN HZ:"
225 PRINT "F C1= ";FC1
230 PRINT
235 PRINT "F C2= ";FC2
240 PRINT
245 PRINT "F CE= ";FCE
250 PRINT
255 PRINT "THE HIGH FREQUENCY 3
DB POINT IS= ";FP
260 PRINT
265 INPUT "TO DO STRAY CAP. ENTE
R 1, TO END ENTER 0.  ":A
270 IF A=1 THEN 280 ELSE 300
275 PRINT
280 INPUT "STRAY CAP. COLLECTOR
OR GROUND OR SUBSRATA CS= ":CS
285 FCS=((1/(RL1*CS))/6.283185)
290 PRINT
295 PRINT "H.F. 3 D.B. DUE TO CS
=";FCS
300 END
```

Run

```
A SIMPLIFIED LOW AND HIGH FREQ.
ANALYSIS FOR THE COMMON EMITTER
AMPLIFIER.

R1=  270000

R2=  62000

RE=  2000

RC=  10000

RL=  6800
```

```
RS=   500

C1=   .000006

C2=   .00001

CE=   .0001

C PIE=  1.5E-11

C MU=  2.E-12

I-C D.C.=  .001475

BETA A.C.=  950

THE LOW FREQUENCY E DB POINT IS
DETERMINED BY THE LARGER OF THE
FOLLOWING THREE FREQUENCIES IN H
Z:
F C1=  2.04706722

F C2=  .9473508981

F CE=  89.61371457

THE HIGH FREQUENCY 3 DB POINT IS
=  663988.3834

TO DO STRAY CAP. ENTER 1, TO END
 ENTER 0

CS=  .0000000001
H.F. 3 D.B. DUE TO CS=
 393206.3492
```

Design of a Common-Source N-JFET Amplifier with a Given DC Operating Point

Circuit Description and Summary of the Analysis

Junction field-effect transistors (J-FET) and metallic-oxide-silicon field-effect transistors (MOSFET) are commonly used in radio-frequency amplifiers and tuners, where a low noise figure is required. They also find applications in low-noise amplifiers for sensors. The MOSFET transistor is used in the large- and medium-scale integrated circuits of micro-computers and in the newer C-MOS technology. In general, they are slow compared to the BJT, but they have the advantage of higher DC and AC input impedances. C-MOS IC's are also found in constant-memory RAM's.

The Common-Source N-JFET amplifier is illustrated in Figure 11.

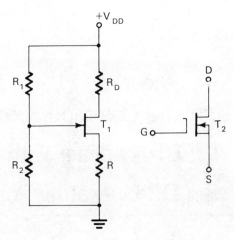

FIGURE 11. A common source N-JFET amplifier.

In this design program, the drain current is modeled by an equation that assumes operation of the transistor in its saturation region or beyond pinch-off. This region is defined by the equation:

$$V_{DS} > V_{GS} - V_P \qquad (1)$$

When equation (1) is satisfied, the drain current may be modeled by means of the following equation:

$$I_D = I_{DSS} * (1 - (V_{GS}/V_P))^2 \qquad (2)$$

This simple nonlinear equation for the drain current in the saturation region is used by the great majority of texts. If equation (1) is not satisfied, the program will so inform the user. In general, a linear amplifier must be designed to operate in the saturation region for good linear amplification.

Sample Program:

For our sample program we elect to design an amplifier using the 2N6451 transistor, using the average specifications for IDSS and VP. According to the specification sheets, the average values are IDSS = 12.5 mA and VP = − 2 volts. The sample program follows.

Enter Program 11.

Program Listing

```
10 PRINT "DESIGN OF A COMMON SOU
RCE NJFET AMPLIFIER FOR A GIVEN
DC OPERATING POINT"
20 PRINT
30 INPUT "VDD= ":VDD
40 INPUT "TARGET VDS= ":VDS
50 INPUT "TARGET ID= ":ID
60 INPUT "IDSS= ":IDSS
70 INPUT "VP= ":VP
80 VGS=(VP*(1-SQR(ID/IDSS)))
90 PRINT "VGS= ";VGS
100 IF VDS<VGS-VP THEN 280
110 PRINT
120 INPUT "CHOOSE R= ":R
130 INPUT "CHOOSE R2= ":R2
140 R1=(((VDD-ID*R)*R2)/(VGS+ID*
R))
150 RD=((VDD-ID*R-VDS)/ID)
160 PRINT
170 PRINT "R1= ";R1
180 PRINT "R2= ";R2
190 PRINT "RD= ";RD
200 PRINT "R= ";R
210 PRINT
220 PRINT "TO REPEAT FOR ANOTHER
 IDSS AND VP, ENTER 1"
230 PRINT "TO END PROGRAM, ENTER
 0"
240 INPUT A
250 IF A=1 THEN 60
260 IF A=0 THEN 290
270 GOTO 290
280 PRINT "TRANSISTOR IS NOT IN
SATURATION. TRY AGAIN."
290 END
```

Run

```
DESIGN OF A COMMON SOURCE AMPLI-
FIER FOR A GIVEN DC OPERATING
POINT
```

```
VDD=  25
TARGET VDS=  3
TARGET ID=  .003
IDSS=  .0125
VP= -2
VGS= -1.020204103

CHOOSE R= 1000
CHOOSE R2=  120000

R1=  1333470.791
R2=  120000
RD=  6333.333333
R=  1000
A=  0
```

Comments: As in any design, the solution is not unique, but dependent on the overall objectives of the designer. In this case I elected to design for a large small-signal voltage gain as well, by producing a design with a large value for the drain resistor, R_D. The user may design for a smaller gain by increasing the voltage across the transistor.

The design may be completed for realization by choosing 5% resistors. These are:

$$R_1 = 1.3 \text{ megohms} \qquad R_2 = 120K \text{ ohms}$$

$$R = 1000 \text{ ohms} \qquad R_D = 6200 \text{ ohms}$$

Analysis of a Common-Source N-JFET or Depletion-Mode MOSFET Amplifier for its DC Operating Point and all Mid-Band Gains

Circuit Description and Summary of the Analysis

The common-source N-JFET amplifier is illustrated in Figure 12. A brief description of the transistor may be found in Program 11.

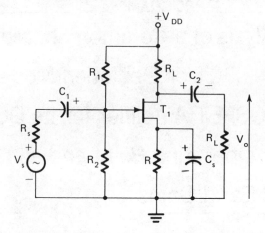

FIGURE 12. A common source N-JFET linear amplifier.

This program repeats many of the steps of Program 11, because the DC operating point must be calculated prior to the analysis for the mid-band gains. While Program 11 is suitable for the beginner who may wish to develop a better understanding of biasing, Program 12 is recommended for the more advanced user. Equations (1) and (2) of Program 11 apply to Program 12 as well. All voltages and currents are defined in the standard manner.

Overall Voltage Gain $A_{Vs} = V_o / V_s$
Circuit Voltage Gain $A_{Vi} = V_O / V_i$
Current Gain $= I_{out} / I_{in}$; $I_{out} = V_o / R_L$, $I_{in} = V_i / R_B$

The equations for the above may be found in the Program LISTING. The mid-band frequency range is again defined in the usual manner. At the lower end it is bounded by the frequency at which capacitors C_1 and C_2 may be neglected, and where capacitor C_s effectively shorts the source to ground. At the high frequency end, it is defined as that frequency below which the effects of the transistor's internal capacitances may be neglected. These may be calculated by using Program 13.

Sample Program:

This example uses the circuit analyzed for its DC operating point by the example of Program 11. The relatively small voltage gains are typical of this type of transistor. However, it should be noted that the input impedance (real at mid-band) is relatively large compared to that in any of the circuits using bipolar-junction transistors. If the particular values for IDSS and VP for the transistor to be used are unknown, the user should RUN this program for the ranges as specified by the manufacturer.

Enter Program 12.

Program Listing

```
10 PRINT "ANALYSIS OF A CS NJFET
 FOR D.C. OP. POINT & ALL MID-BA
ND GAINS"
20 PRINT
30 INPUT "R1=":R1
40 PRINT
50 INPUT "R2=":R2
60 PRINT
70 INPUT "RD=":RD
80 PRINT
90 INPUT "R=":R
100 PRINT
110 INPUT "RS=":RS
120 PRINT
130 INPUT "RL=":RL
140 PRINT
150 INPUT "VDD=":VDD
160 PRINT
170 INPUT "IDSS=":IDSS
180 PRINT
190 INPUT "VP=":VP
200 PRINT
210 REM   VGS^2+A*VGS+B*VP^2=0

220 A=(((VP^2)/(R*IDSS))-2*VP)
230 VG=((VDD*R2)/(R1+R2))
240 B=((1-(VG/(R*IDSS)))*(VP^2))
250 VGS1=((-A+((A^2)-4*B)^(1/2))
/2)
260 VGS2=((-A-((A^2)-4*B)^(1/2))
/2)
270 PRINT "VGS1= ";VGS1
280 PRINT "VGS2= ";VGS2
290 PRINT
300 PRINT "CHOOSE CORRECT VGS.
CORRECT ABS(VGS) < ABS(VP)."
310 INPUT "VGS= ":VGS
320 ID=(IDSS*((1-(VGS/VP))^2))
330 PRINT
340 PRINT "ID= ";ID
```

```
350 PRINT
360 VDS=(VDD-(ID*(R+RD)))
370 PRINT "VDS= ";VDS
380 A=VGS-VP
390 IF VDS<A THEN 400 ELSE 420
400 PRINT "TRANSISTOR IS NOT IN
SATURATION"
410 GOTO 610
420 PRINT
430 GM=-2*IDSS*(1-VGS/VP)/VP
440 AVI=-(GM*RD*RL)/(RD+RL)
450 RB=(R1*R2)/(R1+R2)
460 AVS=(RB/(RB+RS))*AVI
470 AI=AVI*RB/RL
480 AP=AI*AVI
490 PRINT "GM= ";GM
500 PRINT "AVI=";AVI
510 PRINT "AVS=";AVS
520 PRINT "AI= ";AI
530 PRINT "AP= ";AP
540 PRINT
550 PRINT "ZIN=";RB
560 PRINT "ZOUT= ";RD
570 PRINT
580 PRINT "TO REPEAT FOR ANOTHER
 IDSS & VP, ENTER 1. TO END ENTE
R 0"
590 INPUT X
600 IF X=1 THEN 160
610 END
```

Run

```
ANALYSIS OF A CS NJFET FOR D.C.
OP. POINT & ALL MID-BAND GAINS

R1=  1300000
R2=  120000
RD=  6200
R=  1000
RS= 1000
RL= 100000
VDD= 25
IDSS=  .0125
VP= -2

VGS1= -1.001701102
VGS2= -3.318298898

CHOOSE CORRECT VGS. CORRECT ABS(
```

```
VGS) < ABS(VP).
VGS= -1.001701102
ID=  .0031143772
VDS=  2.576484481

GM= .0062393681
AVI=-36.42568955
AVS=-36.09711326
CURRENT GAIN=-40.01695471
POWER GAIN= 1457.645169

ZIN= 109859.1549
ZOUT=  6200

TO REPEAT FOR ANOTHER IDSS & VP
ENTER 1. TO END ENTER 0
X= 0
```

A Simplified Low- and High- Frequency Analysis for the Common-Source J-FET Amplifier

Circuit Description and Summary of the Analysis

This program may be used to calculate the low- and high-frequency 3dB points of the common-source J-FET amplifier illustrated by Figure 12. One of the results of Program 12 will be required as an input: the transconductance, g_m. The user will also be required to input values for the transistor's gate-source and gate-drain capacitors. Typical, maximum, and minimum values can usually be found in the manufacturer's specifications. By repeating both Program 12 and Program 13 for the specification spreads in IDSS and VP, the user can determine the bounds of a circuit's operation.

Since the depletion-mode MOSFET is similar in operation to the J-FET, this transistor type may also be analyzed by this program, though somewhat less exactly.

The high-frequency 3dB point of this circuit is limited by the Miller Effect. For very wide-band applications, the common-base amplifier may be a better choice.

Sample Program:

This example uses the circuit analyzed for its DC operating point and mid-band gains by Program 12. The user should note that all frequencies are calculated in Hz. The low frequency 3dB point is usually controlled by the source by-pass capacitor C_S.

The values for C-GS and C-GD may be obtained from the specification sheets for the 2N6451 transistor. Using the maximum values for the capacitances produces the minimum value for the high-frequency 3dB point which can be realized in practice.

Some comments are in order regarding the calculation of the low-frequency 3dB point. In theory, we should take the low-frequency point as the sum of the three frequencies associated with the three capacitors C_1, C_2, and C_S; however, in good designs the frequencies associated with the coupling capacitors C_1 and C_2 are less than one-tenth of the frequency associated with the source by-pass capacitor C_S, and may be neglected for practical purposes.

Enter Program 13.

Program Listing

```
10 PRINT "A SIMPLIFIED LOW AND H
IGH FREQUENCY ANALYSIS FOR THE C
OMMON SOURCE AMPLIFIER"
15 PRINT
20 INPUT "R1=":R1
25 PRINT
30 INPUT "R2= ":R2
35 PRINT
40 INPUT "RD= ":RD
45 PRINT
50 INPUT "R= ":R
55 PRINT
60 INPUT "RS= ":RS
65 PRINT
70 INPUT "RL= ":RL
75 PRINT
80 INPUT "C1= ":C1
85 PRINT
90 INPUT "C2= ":C2
95 PRINT
```

```
100 INPUT "CS= ":CS
105 IF A=1 THEN 155
110 PRINT
115 INPUT "C-GS= ":CGS
120 PRINT
125 INPUT "C-GD= ":CGD
130 PRINT
135 INPUT "GM AT OP. POINT= ":GM
140 RB=(R1*R2)/(R1+R2)
145 RSP=(RS*RB)/(RS+RB)
150 RLP=(RL*RD)/(RL+RD)
155 WC1=(1/((RS+RB)*C1))
160 WC2=(1/((RD+RL)*C2))
165 RP=R*(1/GM)/(R+(1/GM))
170 WCS=(1/(RP*CS))
175 IF A=1 THEN 190
180 CIN=(CGS+CGD*(1+(GM*RLP)+(RL
P/RSP)))
185 WHIGH=(1/(RSP*CIN))
190 PRINT
195 PRINT "THE LOW FREQUENCY 3 D
B POINT IS THE LARGER OF THE FOL
LOWING THREE FREQUENCIES IN HZ:"
200 PRINT "F-C1= ";WC1/6.283185
205 PRINT "F-C2= ";WC2/6.283185
210 PRINT "F-CS= ";WCS/6.283185
215 IF A=1 THEN 230
220 PRINT
225 PRINT "THE HIGH FREQUENCY 3
DB POINT IS= ";WHIGH/6.283185
230 PRINT
235 PRINT "THIS COMPLETES THE AN
ALYSIS"
240 PRINT "TO REPEAT FOR ANOTHER
 OPERATING POINT, ENTER 2"
245 PRINT "TO REPEAT FOR ANOTHER
 SET OF CAPACITORS, ENTER 1. TO
END ENTER 0"
250 INPUT A
255 IF A=2 THEN 110
260 IF A=1 THEN 75
265 IF A=0 THEN 270
270 END
```

Run

```
A SIMPLIFIED LOW AND HIGH FREQUE
NCY ANALYSIS FOR THE COMMON SOUR
CE AMPLIFIER
```

```
R1=   1300000

R2=   120000

RD=   6200

R=    1000

RS=   1000

RL=   100000

C1=   .000001

C2=   .000004

CS=   .00001

C-GS=  2.5E-11

C-GD=  5.E-12

GM AT OP POINT=  .0062393681

THE LOW FREQUENCY 3 DB POINT IS
THE LARGER OF THE FOLLOWING THRE
E FREQUENCIES IN HZ:
F-C1=  1.435650046
F-C2=  .3746585473
F-CS=  115.2181274

THE HIGH FREQUENCY 3 DB POINT IS
=  664793.3204

TO REPEAT FOR ANOTHER OPERATING
POINT, ENTER 2
TO REPEAT FOR ANOTHER SET OF CAP
ACITORS, ENTER 1. TO END ENTER 0
A=  0
THIS COMPLETES THE ANALYSIS
```

While this example is one of analysis, this program has been written so as to be useful in design. The user may, for example, design for a given low-frequency 3dB point, by changing the input values for the capacitors C_1, C_2, and C_s, by choosing the program branching option labeled number 1. Option number 2 may be used to analyze the circuit for different operating points.

Analysis of a Common-Gate N-JFET or Depletion-Mode MOSFET Amplifier for its DC Operating Point and all Mid-Band Gains

Circuit Description and Summary of the Analysis

The common-base N-JFET amplifier is illustrated in Figure 14. This circuit has a relatively low input impedance and a moderately high output impedance. Since it is not affected by the Miller Effect, its high-frequency performance is superior to that of the common-source amplifier. These characteristics make this circuit very useful as the first stage of the RF section of high-frequency communications systems. The noise figure is also usually less than that of a BJT.

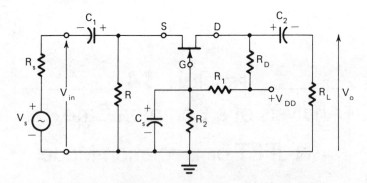

FIGURE 14. A single stage common gate linear amplifier.

This program begins with the calculation of the DC operating point of the circuit. The user must choose the correct gate-source potential from the two choices offered by the computer. Given this choice, the program will then calculate all mid-band gains. These are defined in Program 12. The equations may be found in the Program Listing. Mid-band is defined in the usual manner.

Sample Program:

This example is an illustration of a simple application of the common-base amplifier as a wide-band RF pre-amplifier with an untuned input. The circuit was designed to operate with the 2N5397 N-JFET transistor, for which the specifications may be found in Appendix B. Referring to this specification sheet, the user should note that

$$10 \text{ mA} \le \text{IDSS} \le 30 \text{ mA}.$$

At the 30 mA point, $V_P = -6$ Volts, while $V_P = -1$ Volt at the 10 mA point. These data are required to analyze the circuit for the given components.

The input impedance will range from a high of about 194 ohms to a low of about 97 ohms. In order to match this range of input impedance to an antenna impedance of, say, 52 ohms, an RF transformer with a step-up turns ratio $n \cong \sqrt{3}$ must be used. The turns ratio, n, is defined by the equation $n = N_i / N_o$, where N_i is the number of turns on the input side and N_o is the number of turns on the output side. For an ideal transformer, n may be determined by solving the equation

$$n^2 = Z_{in} / Z_{ant.}$$

Wide-band untuned pre-amplifiers should be used only in locations remote from strong stations, in order to prevent saturation. If there is a strong station nearby, the amplifier must be tuned for best performance.

Enter Program 14.

Program Listing

```
10 PRINT "ANALYSIS OF A CB NJFET
   FOR D.C. OP. POINT & ALL MID-BA
ND GAINS"
20 PRINT
30 INPUT "R1=":R1
40 PRINT
50 INPUT "R2=":R2
60 PRINT
70 INPUT "RD=":RD
80 PRINT
90 INPUT "R=":R
100 PRINT
110 INPUT "RS=":RS
120 PRINT
130 INPUT "RL=":RL
140 PRINT
150 INPUT "VDD=":VDD
160 PRINT
170 INPUT "IDSS=":IDSS
180 PRINT
190 INPUT "VP=":VP
200 PRINT
210 REM   VGS^2+A*VGS+B*VP^2=0

220 A=(((VP^2)/(R*IDSS))-2*VP)
230 VG=((VDD*R2)/(R1+R2))
240 B=((1-(VG/(R*IDSS)))*(VP^2))
250 VGS1=((-A+((A^2)-4*B)^(1/2))
/2)
260 VGS2=((-A-((A^2)-4*B)^(1/2))
/2)
270 PRINT "VGS1= ";VGS1
280 PRINT "VGS2= ";VGS2
290 PRINT
300 PRINT "CHOOSE CORRECT VGS.
CORRECT ABS(VGS)< ABS(VP)."
310 INPUT "VGS= ":VGS
320 ID=(IDSS*((1-(VGS/VP))^2))
330 PRINT
340 PRINT "ID= ";ID
350 PRINT
```

```
360 VDS=(VDD-(ID*(R+RD)))
370 PRINT "VDS= ";VDS
380 A=VGS-VP
390 IF VDS<A THEN 400 ELSE 420
400 PRINT "TRANSISTOR IS NOT IN
SATURATION"
410 GOTO 610
420 PRINT
430 GM=-2*IDSS*(1-VGS/VP)/VP
440 AVI=(GM*RD*RL)/(RD+RL)
450 ZI=(R*(1/GM))/(R+(1/GM))
460 AVS=(AVI*ZI)/(RS+ZI)
470 AI=AVI*ZI/RL
480 AP=AI*AVI
490 PRINT "GM= ";GM
500 PRINT "AVI=";AVI
510 PRINT "AVS=";AVS
520 PRINT "CURRENT GAIN=";AI
530 PRINT "POWER GAIN=";AP
540 PRINT
550 PRINT "ZIN=";ZI
560 PRINT "ZOUT=";RD
570 PRINT
580 PRINT "TO REPEAT FOR ANOTHER
 IDSS & VP, ENTER 1. TO END ENTE
R 0"
590 INPUT X
600 IF X=1 THEN 160
610 END
```

Run

```
ANALYSIS OF A CB NJFET FOR D.C.
OP. POINT & ALL MID-BAND GAINS

R1=  1500000
R2=  150000
RD=  2000
R=  1000
RS= 150
RL= 250000
VDD= 18
IDSS=  .03
VP= -6

VGS1= -3.513961056
VGS2= -9.686038944

CHOOSE CORRECT.VGS. CORRECT ABS(
```

```
VGS) < ABS(VP).
VGS= -3.513961056

ID=  .0051503247
VDS=  2.549025922

GM= .0041433982
AVI= 8.221028254
AVS= 4.640690416
CURRENT GAIN= .0063934604
POWER GAIN= .0525608187

ZIN= 194.4239884
ZOUT= 2000
```

TO REPEAT FOR ANOTHER IDSS & VP __
```
ENTER 1. TO END ENTER 0
X= 1
IDSS=  .01
VP= -1

VGS1= -.5341159397
VGS2= -1.56588406
```

CHOOSE CORRECT VGS. CORRECT ABS(
```
VGS) < ABS(VP).
VGS= -.5341159397

ID=  .0021704796
VDS=  11.48856127

GM= .0093176812
AVI= 18.48746271
AVS= 7.256666687
CURRENT GAIN= .0071672936
POWER GAIN= .1325050738

ZIN= 96.92100192
ZOUT= 2000

TO REPEAT FOR ANOTHER IDSS & VP
ENTER 1. TO END ENTER 0
X= 0
```

A Simplified Low- and High-Frequency Analysis for the Common-Gate Transistor Amplifier

Circuit Description and Summary of the Analysis

This program is designed to be used together with Program 14, which illustrates the circuit and calculates the DC operating point and all mid-band gains. In this circuit, the low-frequency 3dB point is determined by either C_1, C_2, or C_s, while the high-frequency 3dB point is determined by the transistor's internal capacitances, or perhaps by the stray capacitances of the circuit itself. Since the Miller Effect is absent in the common-gate amplifier, the circuit is particularly susceptible to stray-capacitance effects.

The user will be required to input the values of all circuit capacitors, the internal capacitance of the transistor, and the transconductance at the operating point. The program is designed to permit the user to repeat the capacitor inputs. This permits the program to be used for design as well as for the calculation of the spreads in the 3dB points due to transistor-specification spreads.

Sample Program:

This example is a continuation of the example of Program 14, again using the 2N5397 N-JFET transistor. In order to simplify the example, the spread in the transistor's transconductance, g_m, has been averaged:

$$g_m = (g_{m_{max}} - g_{m_{min}})/2 = 5.7 \text{ millimhos}.$$

This will result in an average estimator for the 3dB frequencies $f - C_1$ and $f - CGS$. The other 3dB frequencies are not affected. The values of the internal capacitances of the transistor may be obtained from the manufacturer's specification sheet in Appendix B. The user should note that these are specified in the Hybrid form. Consequently:

$$C_{GS} = C_{iss} - C_{rss} \text{ and } C_{GD} = C_{rss};$$

where C_{iss} and C_{rss} are given by the specification sheet. Using the maximum values specified and neglecting the effect of stray capacitance generates the higher bound for the high-frequency 3dB point.

This is an example of a wide-band voltage amplifier. In the example, two different sets of capacitors result in low-frequency 3dB points of 53 Hz and 950 KHz, respectively. The high frequency 3dB point is about 66.8 MHz.

Program Listing

```
10 PRINT " A SIMPLIFIED LOW AND
HIGH FREQUENCY ANALYSIS FOR THE
COMMON BASE AMPLIFIER"
15 PRINT
20 INPUT "R1=":R1
25 PRINT
30 INPUT "R2= ":R2
35 PRINT
40 INPUT "RD= ":RD
45 PRINT
50 INPUT "R= ":R
55 PRINT
60 INPUT "RS= ":RS
65 PRINT
70 INPUT "RL= ":RL
75 PRINT
80 PRINT
85 INPUT "C1= ":C1
90 PRINT
95 INPUT "C2= ":C2
```

```
100 PRINT
105 INPUT "CS= ":CS
110 IF X=1 THEN 145
115 PRINT
120 INPUT "C-GS= ":CGS
125 PRINT
130 INPUT "C-GD= ":CGD
135 PRINT
140 INPUT "GM AT OP. POINT= ":GM
145 RB=(R1*R2)/(R1+R2)
150 RP=R*(1/GM)/(R+(1/GM))
155 RLP=(RL*RD)/(RL+RD)
160 WC1=(1/((RS+RP)*C1))
165 WC2=(1/((RD+RL)*C2))
170 WCS=(1/(RB*CS))
175 WGD=(1/(RLP*CGD))
180 WGS=(((1/RS)+(1/R)+GM)/CGS)
185 PRINT
190 PRINT "THE LOW FREQUENCY 3 D
B POINT IS THE LARGER OF THE FOL
LOWING THREE FREQUENCIES IN HZ:"
195 PRINT "F-C1= ";WC1/6.283185
200 PRINT "F-C2= ";WC2/6.283185
205 PRINT "F-CS= ";WCS/6.283185
210 IF X=1 THEN 240
215 PRINT
220 PRINT "THE HIGH FREQUENCY 3
DB POINT IS THE SMALLER OF THE F
OLLOWING TWO FREQUENCIES:"
225 PRINT
230 PRINT "F-CGD= ";WGD/6.283185
235 PRINT "F-CGS= ";WGS/6.283185
240 PRINT
245 PRINT "TO REPEAT FOR ANOTHER
 OPERATING POINT, ENTER 2"
250 PRINT "TO REPEAT FOR ANOTHER
 SET OF CAPACITORS ENTER 1. TO E
ND ENTER 0."
255 INPUT X
260 IF X=2 THEN 115
265 IF X=1 THEN 80
270 PRINT "THIS COMPLETES THE AN
ALYSIS"
275 END
```

Run

```
A SIMPLIFIED LOW AND HIGH FREQUE
NCY ANALYSIS FOR THE COMMON BASE
 AMPLIFIER
```

```
R1=  1500000

R2=  150000

RD=  2000

R=  1000

RS=  150

RL=  250000

C1=  .00001

C2=  .000002

CS=  .000001

C-GS=  3.8E-12

C-GD=  1.2E-12

GM AT OP,POINT=  .0057

THE LOW FREQUENCY 3 DB POINT IS
THE LARGER OF THE FOLLOWING THRE
E FREQUENCIES IN HZ:
F-C1=  53.18394867
F-C2=  .3157836327
F-CS=  1.167136306

THE HIGH FREQUENCY 3 DB POINT IS
 THE SMALLER OF THE FOLLOWING TW
O FREQUENCIES:
F-CGD=  66845079.37
F-CGD=  559834520.2
C1=  5.6E-10

C2=  .000000001

CS=  .000000001

THE LOW FREQUENCY 3 DB POINT IS
THE LARGER OF THE FOLLOWING THRE
E FREQUENCIES IN HZ:
F-C1=  949713.3691
F-C2=  631.5672654
F-CS=  1167.136306

THIS COMPLETES THE ANALYSIS
```

_____ PROGRAM **16** _____

Solution of the Matrix Equation
$V = Z \times I$ by Means of the
Modified Gauss-Jordan
Algorithm

Circuit Description and Summary of the Analysis

This program may be used to solve a set of Mesh Equations for a circuit containing resistors, inductors and capacitors. Active elements may be included provided they are modeled by linear theory. Mathematically speaking, this program uses a modified version of the Gauss-Jordan algorithm to solve a set of linear equations with complex coefficients. This problem will be familiar to electrical engineers.

The user will require enough understanding of circuit theory to construct the N original linear equations from the schematic. A general example of such a circuit diagram is illustrated in Figure 16a for a four mesh circuit.

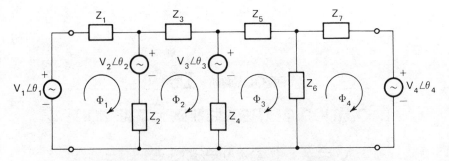

FIGURE 16a. A general example of a four mesh circuit.

The corresponding matrix equation is given by:

$$
\begin{bmatrix}
V_1 \; \underline{/\theta}_1 - V_2 \; \underline{/\theta}_2 \\[4pt]
V_2 \; \underline{/\theta}_2 - V_3 \; \underline{/\theta}_3 \\[4pt]
V_3 \; \underline{/\theta}_3 \\[4pt]
- V_4 \; \underline{/\theta}_4
\end{bmatrix}
=
$$

$$
\begin{bmatrix}
(Z_1 + Z_2) & -Z_2 & 0 & 0 \\[6pt]
-Z_2 & (Z_2 + Z_3 + Z_4) & -Z_4 & 0 \\[6pt]
0 & -Z_4 & (Z_4 + Z_5 + Z_6) & -Z_6 \\[6pt]
0 & 0 & -Z_6 & (Z_6 + Z_7)
\end{bmatrix}
\mathbf{X}
\begin{bmatrix}
I_1 \\[4pt] I_2 \\[4pt] I_3 \\[4pt] I_4
\end{bmatrix}
$$

where, in general,

$$Z_i = R_i + j(\omega L_i - 1/\omega C_i)$$

When entering the matrices V and Z, the user must input the real and imaginary parts of each component. While this is straightforward for the Z matrix, Euler's identities must be used for the voltage phasors. For example, for the first component of the voltage vector, we may write:

$$V_1 \; \underline{/\theta}_1 - V_2 \; \underline{/\theta}_2 = \text{Real Part} + \text{Imaginary Part}$$

where:

$$\text{Real Part} = V_1 \cos\Theta_1 - V_2 \cos\Theta_2$$
$$\text{Imaginary Part} = V_1 \sin\Theta_1 - V_2 \sin\Theta_2$$

The program is dimensioned in Line 20 to accept up to a 9-by-9 matrix, that is, 9 equations. The limit may be increased by the user. As is the case with any large numerical program, the accuracy of the solutions for the currents will depend in part on the number of equations, and in part on the numerical precision of the particular micro-computer used. This will be most evident when some of the solutions for the currents should be identically zero. For a matrix as large as 9-by-9, the user would be wise to limit the number of significant figures in the solution to about 4 unless double precision is used.

Sample Program:

The sample program is an example of a simple three-mesh circuit containing resistors, inductors, and capacitors whose values have been chosen so as to simplify the inputs. The circuit is illustrated by Figure 16b. The two voltage sources have an identical frequency of $\omega = 10$ Radians per second, or $10/2\pi$ Hz. If the phasor notation for the voltage sources is taken to indicate their RMS values, then the solutions for the currents will be RMS values in amperes.

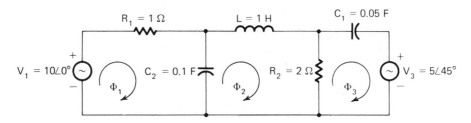

FIGURE 16b. The three mesh circuit used for the sample.

The corresponding matrix equation is given by:

$$
\begin{bmatrix}
10 + j0 \\
0 + j0 \\
-5/2 - j5/2
\end{bmatrix}
=
\begin{bmatrix}
(1 - j1) & -j1 & 0 \\
-j1 & (2 + j9) & -2 \\
0 & -2 & (2 - j2)
\end{bmatrix}
\times
\begin{bmatrix}
I_1 \\
I_2 \\
I_3
\end{bmatrix}
$$

Enter Program 16.

Program Listing

```
10 PRINT "SOLUTION OF THE MATRIX
   EQUATION V=Z*I BY WAY OF THE MO
DIFIED GAUSS-JORDAN ALGORITHM"
15 PRINT
20 DIM R(9,10),X(9,10)
25 INPUT "NO OF EQUATIONS?":N
30 NP=N+1
35 PRINT
40 PRINT "ENTER THE Z MATRIX SEP
ARATING THE REAL AND IMAGINARY P
ARTS BY A COMMA:"
45 FOR A=1 TO N STEP 1
50 FOR B=1 TO N STEP 1
55 PRINT
60 PRINT "ENTER R(";A;",";B;"),X
(";A;",";B;")"
65 INPUT R(A,B),X(A,B)
70 NEXT B
75 NEXT A
80 PRINT
85 PRINT "ENTER THE V MATRIX"
90 PRINT
95 FOR A=1 TO N STEP 1
100 PRINT "REAL V(";A;"),IMG V("
;A;")"
105 INPUT R(A,NP),X(A,NP)
110 NEXT A
115 FOR I=1 TO N STEP 1
120 ALPAR=R(I,I)
125 ALPAI=X(I,I)
130 ALPA=(ALPAR^2+ALPAI^2)
135 IF ALPA=0 THEN 260
140 FOR J=I TO NP STEP 1
145 R(I,J)=(R(I,J)*ALPAR+X(I,J)*
ALPAI)/ALPA
150 X(I,J)=X(I,J)*ALPAR/ALPA-R(I
,J)*ALPAI/ALPAR+X(I,J)*(ALPAI^2)
/(ALPA*ALPAR)
155 NEXT J
160 FOR K=1 TO N STEP 1
165 IF (K-I)=0 THEN 200
170 BETAR=R(K,I)
175 BETAI=X(K,I)
180 FOR J=I TO NP STEP 1
185 R(K,J)=R(K,J)-(BETAR*R(I,J)-
BETAI*X(I,J))
```

```
190 X(K,J)=X(K,J)-(BETAR*X(I,J)+
BETAI*R(I,J))
195 NEXT J
200 NEXT K
205 NEXT I
210 PRINT
215 PRINT
220 PRINT "THE MESH CURRENTS ARE
:"
225 PRINT
230 FOR M=1 TO N STEP 1
235 PRINT "REAL I(";M;")=";R(M,N
P)
240 PRINT "IMG  I(";M;")=";X(M,N
P)
245 PRINT
250 NEXT M
255 GOTO 265
260 PRINT "THE MAIN DIAGONAL ELE
MENT IS ZERO"
265 END
```

Run

```
SOLUTION OF THE MATRIX EQUATION
 V=Z*I BY WAY OF THE MODIFIED
 GAUSS-JORDAN ALGORITHM

N=  3


ENTER THE Z MATRIX SEPARATING TH
E REAL AND IMAGINARY PARTS BY A
COMMA:


R( 1 , 1 )= 1
X( 1 , 1 )=-1

R( 1 , 2 )= 0
X( 1 , 2 )=-1

R( 1 , 3 )= 0
X( 1 , 3 )= 0

R( 2 , 1 )= 0
X( 2 , 1 )=-1

R( 2 , 2 )= 2
X( 2 , 2 )= 9
```

```
R( 2 , 3 )=-2
X( 2 , 3 )= 0

R( 3 , 1 )= 0
X( 3 , 1 )= 0

R( 3 , 2 )=-2
X( 3 , 2 )= 0

R( 3 , 3 )= 2
X( 3 , 3 )=-2

ENTER THE V MATRIX

REAL V( 1 )= 10
IMG  V( 1 )= 0

REAL V( 2 )= 0
IMG  V( 2 )= 0

REAL V( 3 )=-3.535533906
IMG  V( 3 )=-3.535533906


THE MESH CURRENTS ARE:
REAL I( 1 )= 4.666814356
IMG  I( 1 )= 4.733229951

REAL I( 2 )= .0664155946
IMG  I( 2 )= .5999556932

REAL I( 3 )=-.2667700493
IMG  I( 3 )=-1.434581309
```

Solution of the Matrix Equation I = Y x V by Means of the Modified Gauss-Jordan Algorithm

Circuit Description and Summary of the Analysis

This program may be used to solve a set of Node Equations for a circuit containing resistors, inductors, and capacitors. Active elements may be included provided they are modeled by linear theory. Mathematically speaking, this program uses a modified version of the Gauss-Jordan Algorithm to solve a set of linear equations with complex coefficients. This problem will be familiar to electrical engineers.

The user will require enough understanding of circuit theory to construct the N original linear equations from the schematic. A general example of a three-node circuit is illustrated in Figure 17a.

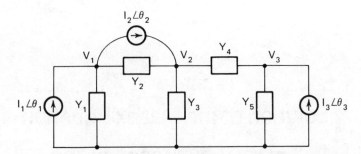

FIGURE 17a. A general example of a three node circuit.

The corresponding matrix equation is given by:

$$\begin{bmatrix} I_1 \underline{/\theta}_1 - I_2 \underline{/\theta}_2 \\ I_2 \underline{/\theta}_2 \\ I_3 \underline{/\theta}_3 \end{bmatrix} = \begin{bmatrix} (Y_1 + Y_2) & -Y_2 & 0 \\ -Y_2 & (Y_2 + Y_3 + Y_4) & -Y_4 \\ 0 & -Y_4 & (Y_4 + Y_5) \end{bmatrix} \times \begin{bmatrix} V_1 \\ V_2 \\ V_3 \end{bmatrix}$$

where, in general,

$$Y_i = 1/R_i + j(\omega C_i - 1/\omega L_i)$$

When entering the matrices I and Y, the user must input the real and imaginary parts of each component. While this is straightforward for the Y matrix, Euler's identities must be used for the current phasors. For example, the first component of the current vector may be written as:

$$I_1 \underline{/\theta}_1 - I_2 \underline{/\theta}_2 = \text{Real Part} + \text{Imaginary Part}$$

where:

Real Part $= I_1 \cos \theta_1 - I_2 \cos \theta_2$
Imaginary Part $= I_1 \sin \theta_1 - I_2 \sin \theta_2$

The program is dimensioned in Line 15 to accept up to 9-by-9 matrix, that is, 9 equations. The limit may be increased by the user. Before increasing the matrix size, the user should read the comments on Program 16.

Sample Program:

The sample program is an example of a simple three-node circuit containing resistors, inductors, and capacitors whose values have been chosen so as to

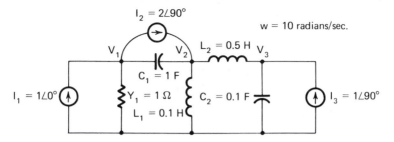

FIGURE 17b. The three node circuit used for the sample.

simplify the inputs. The circuit is illustrated by Figure 17b. The three current sources have identical frequencies of $\omega = 10$ Radians per second, or $10/2\pi$ Hz. If the phasor notation for the current sources is taken to indicate their RMS values, then the solutions for the voltages will be RMS values in volts.

The corresponding matrix equation is given by:

$$\begin{bmatrix} 1 - j2 \\ j2 \\ j1 \end{bmatrix} = \begin{bmatrix} (1 + j10) & -j10 & 0 \\ -j10 & j8.8 & -j0.2 \\ 0 & -j0.2 & j0.8 \end{bmatrix} \times \begin{bmatrix} V_1 \\ V_2 \\ V_3 \end{bmatrix}$$

Enter Program 17.

Program Listing

```
10 PRINT "SOLUTION OF THE MATRIX
   EQUATION I=Y*V BY WAY OF THE MO
   DIFIED GAUSS-JORDAN ALGORITHM"
15 DIM G(9,10),B(9,10)
20 PRINT
25 INPUT "NO OF EQUATIONS?":N
30 PRINT
35 NP=N+1
40 PRINT "ENTER THE Y MATRIX SEP
   ARATING THE REAL AND IMAGINARY P
   ARTS BY A COMMA:"
45 PRINT
50 FOR A=1 TO N STEP 1
55 FOR C=1 TO N STEP 1
60 PRINT "ENTER G(";A;",";C;"),B
   (";A;",";C;")"
65 INPUT G(A,C),B(A,C)
70 NEXT C
```

```
75 NEXT A
80 PRINT
85 PRINT "ENTER THE I MATRIX"
90 PRINT
95 FOR A=1 TO N STEP 1
100 PRINT "REAL I(";A;"),IMG I("
;A;")"
105 INPUT G(A,NP),B(A,NP)
110 PRINT
115 NEXT A
120 FOR I=1 TO N STEP 1
125 ALPAR=G(I,I)
130 ALPAI=B(I,I)
135 ALPA=(ALPAR^2+ALPAI^2)
140 IF ALPA=0 THEN 260
145 FOR J=I TO NP STEP 1
150 G(I,J)=(G(I,J)*ALPAR+B(I,J)*
ALPAI)/ALPA
155 B(I,J)=B(I,J)*ALPAR/ALPA-G(I
,J)*ALPAI/ALPAR+B(I,J)*(ALPAI^2)
/(ALPA*ALPAR)
160 NEXT J
165 FOR K=1 TO N STEP 1
170 IF (K-I)=0 THEN 205
175 BETAR=G(K,I)
180 BETAI=B(K,I)
185 FOR J=I TO NP STEP 1
190 G(K,J)=G(K,J)-(BETAR*G(I,J)-
BETAI*B(I,J))
195 B(K,J)=B(K,J)-(BETAR*B(I,J)+
BETAI*G(I,J))
200 NEXT J
205 NEXT K
210 NEXT I
215 PRINT
220 PRINT "THE NODE VOLTAGES ARE
:"
225 PRINT
230 FOR M=1 TO N STEP 1
235 PRINT "REAL V(";M;")=";G(M,N
P)
240 PRINT "IMG  V(";M;")=";B(M,N
P)
245 PRINT
250 NEXT M
255 GOTO 265
260 PRINT "THE MAIN DIAGONAL IS
ZERO"
265 END
```

Run

SOLUTION OF THE MATRIX EQUATION
 I=Y*V BY WAY OF THE MODIFIED
 GAUSS-JORDAN ALGORITHM

N= 3

ENTER THE Y MATRIX SEPARATING TH
E REAL AND IMAGINARY PARTS BY A
COMMA:

G(1 , 1)= 1
B(1 , 1)= 10

G(1 , 2)= 0
B(1 , 2)=-10

G(1 , 3)= 0
B(1 , 3)= 0

G(2 , 1)= 0
B(2 , 1)=-10

G(2 , 2)= 0
B(2 , 2)= 8.8

G(2 , 3)= 0
B(2 , 3)=-.2

G(3 , 1)= 0
B(3 , 1)= 0

G(3 , 2)= 0
B(3 , 2)=-.2

G(3 , 3)= 0
B(3 , 3)= .8

ENTER THE I MATRIX

REAL I(1)= 1
IMG I(1)=-2

REAL I(2)= 0
IMG I(2)= 2

REAL I(3)= 0
IMG I(3)= 1

```
THE NODE VOLATGES ARE:
REAL V( 1 )= .0604026846
IMG  I( 1 )= .6577181208

REAL V( 2 )= .3261744966
IMG  I( 2 )= .7516778523

REAL V( 3 )= 1.331543624
IMG  I( 3 )= .1879194631
```

Design of Active Low- and High-pass Filters by Means of the Butterworth Polynomials

Circuit Description and Summary of the Analysis

This program is an excellent introduction to the use of integrated circuits in design. The program may be used to design a low- or a high-pass filter of the 1st, 2nd, 3rd, or 4th order. With a simple change in the circuit (R and C interchanged in all of the Figures), it may be used to design high-pass filters.

The computer program will result in a design that is 3dB down at the design frequency and has a "drop-off" that increases with increasing order. For example, a 1st-order filter will be 20dB down at ten times the design frequency, while a 4th-order filter will be more than 20dB down at just twice the design frequency. For further details, the user is referred to reference 1. In any design, the user must be sure to choose an integrated operational amplifier with a high-frequency response exceeding that required for the design.

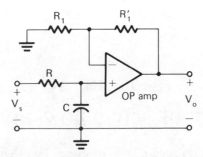

FIGURE 18a. A first order low pass filter using an operational amplifier.

1. 1st-Order Filter Design

A schematic of a typical 1st-order filter (without power supply connections) is illustrated by Figure 18a.

To use this design, the user must choose the 1st-order design in the program, and follow the instructions. To design a high-pass filter, simply interchange R and C. The user should choose the integrated operational amplifier carefully. The gain-bandwidth product of the OP amplifier must be large enough to ensure success of the design. For example, if the gain of the 1st-order filter is taken as 10, and the cut-off frequency is taken as 10 kHz, then the gain-bandwidth product of the OP amplifier must be about $10 \times 10 \times 10 \times 10^3$ to insure "good" design. This will ensure that the 3dB point of the OP amplifier itself is well above the designed 3dB point of the filter. In high-pass filter design, the gain-bandwidth product of the OP amplifier will limit the high-frequency response at the high end. The user should also avoid unusually high gains, as these may produce oscillations.

2. 2nd-Order Filter Design

A schematic of a typical 2nd-order filter (without power supply connections) is illustrated by Figure 18b.

To use this design, the user must choose the 2nd-order design in the program, and follow the instructions. To design a high-pass filter, simply interchange R and C. Be sure to read the comments on the 1st-order design before building the circuit.

3. 3rd-Order Filter Design

A schematic of a typical 3rd-order filter (without power supply connections) is illustrated by Figure 18c.

To use this design, the user must choose the 3rd-order design in the program, and follow the instructions. To design a high-pass filter, simply interchange R and C everywhere in the circuit of Figure 18c. Be sure to read the comments on the 1st-order design before building the circuit.

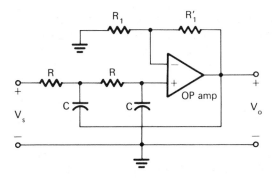

FIGURE 18b. A second order low pass filter using an operational amplifier.

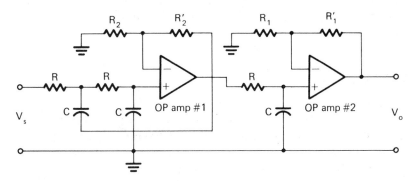

FIGURE 18c. A third order low pass filter using two operational amplifiers.

4. 4th-Order Filter Design

A schematic of a typical 4th-order filter (without power supply connections) is illustrated by Figure 18d.

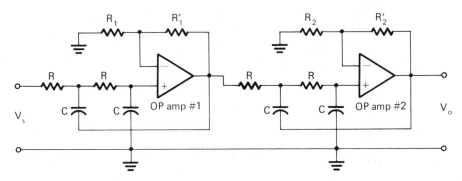

FIGURE 18d. A fourth order low pass filter using two operational amplifiers.

To use this design, the user must choose the 4th-order design in the program, and follow the instructions. To design a high-pass filter, simply interchange R and C everywhere in the circuit of Figure 18d. Be sure to read the comments in the 1st-order design before building the circuit.

Additional Comments:

Because of design constraints dictated by the Butterworth criterion, the overall gain of the odd-order filters can be designed to be greater than that of the even-order filters. If gain is desired, the user may elect to use the 3rd-order filter design. This filter has a 20dB attenuation at a frequency that is just a bit greater than the designed 3dB frequency, while the overall gain may be controlled by choosing a large gain, of say 10 or so, for the 1st-order component.

Sample Program:

Our sample for this program will be two 2nd-order filter designs with low-pass cut-off frequencies of 3,000 Hz and 2,500 Hz, respectively.

Enter Program 18.

Program Listing

```
10 PI=3.141592654
15 PRINT "DESIGN OF ACTIVE LOW
PASS FILTERS BY WAY OF THE BUTTE
RWORTH POLYNOMIALS"
20 PRINT
25 PRINT "DESIGNER MUST SPECIFY
FILTER ORDER:1,2,3,OR 4"
30 INPUT "ORDER ?":N
35 IF N=1 THEN 100
40 IF N=2 THEN 200
45 IF N=3 THEN 300
50 IF N=4 THEN 400
55 PRINT "INPUT ERROR. TRY AGAIN
"
60 GOTO 25
100 REM  FIRST ORDER FILTER DESI
GN
```

Reference: Millman, J. and Halkias, C.C., "Integrated Electronics: Analog and Digital Circuits and Systems", McGraw-Hill Book Company, 1972.

```
102 PRINT
105 PRINT "FIRST ORDER FILTER
DESIGN. REFER TO FIGURE 18A."
110 PRINT
115 INPUT "ENTER 3 DB HIGH FREQU
ENCY LIMIT= ":F
120 INPUT "SPECIFY RESISTOR R= "
:R
125 C=1/(2*PI*F*R)
135 PRINT "C= ";C
140 INPUT "SPECIFY GAIN AV= ":AV
145 INPUT "SPECIFY RESISTOR R1=
":R1
150 R11=R1*(AV-1)
155 PRINT "R1'= ";R11
165 PRINT "THIS COMPLETS YOUR DE
SIGN"
170 PRINT
175 PRINT "TO REPEAT FOR ANY ORD
ER, ENTER 1. TO END ENTER 0."
180 INPUT B
185 IF B=1 THEN 25
190 END
200 REM  SECOND ORDER FILTER DES
IGN
202 PRINT
205 PRINT "SECOND ORDER FILTER
DESIGN. REFER TO FIGURE 18B."
210 PRINT
215 INPUT "ENTER 3 DB HIGH FREQU
ENCY LIMIT= ":F
220 INPUT "SPECIFY RESISTOR R= "
:R
225 C=1/(2*PI*F*R)
230 PRINT "C= ";C
235 PRINT "THE GAIN AV=1.586"
240 INPUT "SPECIFY RESISTOR R1=
":R1
245 R11=.586*R1
250 PRINT "R1'= ";R11
255 PRINT "THIS COMPLETES YOUR D
ESIGN"
260 PRINT
265 PRINT "TO REPEAT FOR ANY ORD
ER DESIGN, ENTER 1. TO END ENTER
 0."
270 INPUT B
275 IF B=1 THEN 25
280 END
300 REM  THIRD ORDER FILTER DESI
```

```
GN
302 PRINT
305 PRINT "THIRD ORDER FILTER
DESIGN. REFER TO FIGURE 18C."
310 PRINT
315 INPUT "ENTER 3 DB HIGH FREQU
ENCY LIMIT= ":F
317 INPUT "SPECIFY RESISTOR R= "
:R
320 C=1/(2*PI*F*R)
325 PRINT "C= ";C
330 PRINT "THE INCIRCUIT GAIN OF
 THE 2ND ORDER OP AMP IS=2."
335 INPUT "SPECIFY RESISTOR R2=
":R2
340 PRINT "R2'= ";R2
345 PRINT "THE CHOICE FOR THE IN
CIRCUIT GAIN OF THE 1ST ORDER OP
 AMP IS YOURS."
350 INPUT "AV1= ":AV1
355 INPUT "SPECIFY RESISTOR R1=
":R1
360 R11=R1*(AV1-1)
365 PRINT "R1'= ";R11
370 PRINT "THIS COMPLETES YOUR D
ESIGN"
375 PRINT
380 PRINT "TO REPEAT FOR ANY ORD
ER, ENTER 1. TO END ENTER 0."
385 INPUT B
390 IF B=1 THEN 25
400 REM  FOURTH ORDER FILTER DES
IGN
402 PRINT
405 PRINT "FOURTH ORDER FILTER
DESIGN. REFER TO FIGURE 18D."
410 PRINT
420 INPUT "ENTER 3 DB HIGH FREQU
ENCY LIMIT= ":F
425 INPUT "SPECIFY RESISTOR R= "
:R
430 C=1/(2*PI*F*R)
435 PRINT "C= ";C
440 PRINT "THE INCIRCUIT GAIN OF
 THE 1ST SECOND ORDER ELEMENT IS
 2.235"
445 INPUT "SPECIFY RESISTOR R1 F
OR 1ST FILTER. R1= ":R1
450 R11=1.235*R1
455 PRINT "R1'= ";R11
```

```
460 PRINT
465 PRINT "THE INCIRCUIT GAIN OF
 THE 2ND SECOND ORDER ELEMENT IS
 1.152."
470 INPUT "SPECIFY RESISTOR R2 F
OR THE 2ND FILTER. R2= ":R2
475 R22=.152*R2
480 PRINT "R2'= ";R22
485 PRINT
490 PRINT "TO REPEAT FOR ANY ORD
ER, ENTER 1. TO END ENTER 0."
495 INPUT B
500 IF B=1 THEN 25
505 END
```

Run

```
DESIGN OF ACTIVE LOW PASS FILTER
S BY WAY OF THE BUTTERWORTH POLY
NOMIALS

SECOND ORDER FILTER DESIGN. REFE
R TO FIGURE 18B.

3 DB HIGH FREQUENCY LIMIT=  3000
RESISTOR R=  1000
C=  5.30516E-08

THE GAIN AV=1.586

RESISTOR R1=  1000
R1'=  586

THIS COMPLETES YOUR DESIGN


SECOND ORDER FILTER DESIGN. REFE
R TO FIGURE 18B.

3 DB HIGH FREQUENCY LIMIT=  2500
RESISTOR R=  1000
C=  6.3662E-08

THE GAIN AV=1.586

RESISTOR R1=  1000
R1'=  586

THIS COMPLETES YOUR DESIGN
```

Noise Figure of a Communications Receiver

Circuit Description and Summary of the Analysis

The "noise figure", F, of a communications receiver, or of any other system, is a convenient way of relating the signal-to-noise ratio of the input to the signal-to-noise ratio of the output, by way of the equation:

$$\left(\frac{S}{N} \right)_{Output} = \frac{1}{F} \times \left(\frac{S}{N} \right)_{Input} \tag{1}$$

with $F \geq 1$.

In this definition, F is a number; however, a decibel definition is usually used,

$$F_{dB} = 10 \log_{10} F \tag{2}$$

We may also use the noise figure to define an effective noise temperature for the system, T_{eff}, in degrees Kelvin:

$$T_{eff} = 290 \, (F - 1). \tag{3}$$

The "290 degree Kelvin" reference (equal to 17 degrees Centigrade or 62.6 degrees Fahrenheit) is a standard used in communications.

105

Program 19 was written so as to enable the user to easily calculate F, F_{dB}, and T_{eff}, for a system of up to four individual units. Since the first unit, or "front end", of a receiver is usually the one which introduces the most noise into the system, calculating F for the first four units will suffice for most appalications. The system is illustrated by Figure 19.

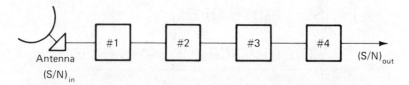

FIGURE 19. Block diagram of the first four stages of a communications receiver.

To use this program, the user must specify the noise figure in dB and the voltage gain in dB of each unit. This information may be obtained from manufacturer's specifications, or by means of experimentation. This procedure is straightforward unless one of the units is a passive component with loss, such as a coaxial cable or a waveguide. Since such cables or guides are a part of nearly all communications systems, it is best to consider this case separately.

A cable may be entered by characterizing this unit as a simple attenuator with some given loss in dB. The loss may be obtained from the manufacturer's specifications, the frequency of operation, and the cable length. For a cable or guide at "room" temperature, we obtain:

$$F_{dB} = \text{loss in dB} \quad ; \quad \text{gain in dB} = -F_{dB} \tag{4}$$

For a cable or guide at some (usually lower) temperature, we obtain:

$$F_{dB} = 1 + \frac{(L-1)T_c}{290} \quad , \tag{5}$$

where T_c = cable or guide temperature in degrees Kelvin.

In all cases where low noise figures are required, a long cable connection should be avoided between the antenna and the first radio-frequency amplifier. In fact, all low-noise-figure systems place the first amplifier directly on the antenna itself. This will be illustrated in the examples.

Sample Programs:

The samples have been selected to illustrate the effect that changing the location of the coaxial cable has on the noise figure of the communications system.

Example 1

Unit # 1: Coaxial Cable with an attenuation of 1dB.

Unit # 2: RF Amplifier using FET's. F_{dB} = 6dB. Gain = 20dB.

Unit # 3: 1st converter. F_{dB} = 10dB. Gain = 6dB.

Unit # 4: 1st IF Amplifier. F_{dB} = 6dB. Gain = 50dB.

Enter Program 19.

Program Listing

```
10 PRINT "NOISE FIGURE OF A COMM
UNICATIONS RECEIVER"
15 PRINT
20 INPUT "NUMBER OF STAGES: 2 TO
 4. N= ":N
25 PRINT
30 PRINT "IF A DEVICE IS AN ATTE
NUATOR SUCH AS A CABLE, ENTER IT
S GAIN AS A NEGATIVE NUMBER IN D
B."
35 PRINT
40 INPUT "ENTER NOISE FIGURE OF
1ST STAGE IN DB. F1= ":F1
45 F11=(10^(F1/10))
50 INPUT "ENTER GAIN OF 1ST STAG
E IN DB. G1= ":G1
55 G11=(10^(G1/10))
60 PRINT
65 INPUT "ENTER NOISE FIGURE OF
2ND STAGE IN DB. F2= ":F2
70 F22=(10^(F2/10))
75 IF N=2 THEN 140
80 INPUT "ENTER GAIN OF 2ND STAG
E IN DB. G2= ":G2
85 G22=(10^(G2/10))
90 PRINT
95 INPUT "ENTER NOISE FIGURE OF
THIRD STAGE IN DB. F3= ":F3
100 F33=(10^(F3/10))
105 IF N=3 THEN 160
110 INPUT "ENTER GAIN OF THIRD S
TAGE IN DB. G3= ":G3
115 G33=(10^(G3/10))
120 PRINT
125 INPUT "ENTER NOISE FIGURE OF
 FOURTH STAGE IN DB. F4= ":F4
130 F44=(10^(F4/10))
135 IF N=4 THEN 180
140 REM  TWO STAGE ANALYSIS
```

```
145 PRINT
150 F=F11+(F22-1)/G11
155 GOTO 195
160 REM   THREE STAGE ANALYSIS

165 PRINT
170 F=F11+((F22-1)/G11)+((F33-1)
/(G11*G22))
175 GOTO 195
180 REM   FOUR STAGE ANALYSIS
185 PRINT
190 F=F11+((F22-1)/G11)+((F33-1)
/(G11*G22))+((F44-1)/(G11*G22*G3
3))
195 REM   PRINT THE RESULTS
200 PRINT "NOISE FIGURE= ";F
205 A=10*(LOG(F)/LOG(10))
210 PRINT
215 PRINT "NOISE FIGURE IN DB= "
;A
220 PRINT
225 TE=290*(F-1)
230 PRINT "EFFECTIVE NOISE TEMP.
 IN DEG. KELVIN= ";TE
235 END
```

Run

```
NOISE FIGURE OF A COMMUNICATIONS
 RECEIVER

NUMBER OF STAGES=  4


F1=  1
G1= -1

F2=  6
G2=  20

F3=  10
G3=  6

F4=  6


NOISE FIGURE=  5.1346026

NOISE FIGURE IN DB=  7.105068364

EFFECTIVE NOISE TEMP. IN DEG. KE
LVIN=  1199.034754
```

Example 2

For the second example, interchange units 1 and 2, placing the FET RF amplifier at the antenna. The cable then is used to connect the RF amplifier with the remainder of the receiving system located indoors.

Run

```
NOISE FIGURE OF A COMMUNICATIONS
 RECEIVER

NUMBER OF STAGES=  4


F1=  6
G1=  20

F2=  1
G2= -1

F3=  10
G3=  6

F4=  6


NOISE FIGURE=  4.106391223

NOISE FIGURE IN DB=  6.13460323

EFFECTIVE NOISE TEMP. IN DEG. KE
LVIN=  900.8534547
```

The two examples demonstrate that by simply moving the RF amplifier to the antenna, we have reduced the noise figure by 1dB, or some 200 degrees in effective noise temperature.

Class A Power Amplifier Design

Circuit Description and Summary of the Analysis

The Class A transistor amplifier is a common circuit that may be found in systems ranging from audio amplifiers to the RF amplifiers of single-side-band communications systems. This program is based on a design algorithm developed by the author, and in this case applied to the case of an emitter-follower class A power amplifier. The emitter-follower design enables the user to match low-impedance loads such as loudspeakers without the use of a transformer. The program was written so as to interact with the user to produce a circuit that will match a user-specified load impedance. The circuit is illustrated in Figure 20.

In addition to designing a circuit to produce a target output impedance (using the mid-band equivalent circuit for the transistor), the program also calculates: the circuit input impedance, ZIN; the circuit voltage gain, AVI; and the circuit power gain, AP. The user is required to input the desired DC operating point and the usual transistor data. In choosing the operating point, the user should note that it is unwise to attempt to obtain an AC power output that exceeds about 1/3 of the DC power input. Attempting to achieve the theoretical limit, one-half of the DC power input, may result in excessive

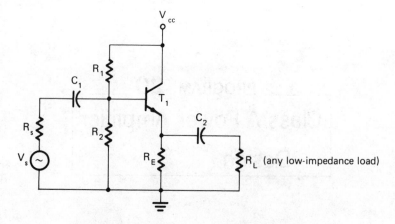

FIGURE 20. A class A emitter follower power amplifier.

distortion because of the nonlinearity of the transistor's characteristics near cut-off and saturation.

Because this program designs a power amplifier, some consideration must be given to the actual temperature of the junction of the transistor. The user must input the ambient temperature, and the thermal resistance from the junction of the transistor to ambient. This information may be obtained from the specification sheets for the transistor. At high power levels, a heat sink will be required.

When a heat sink is used, the thermal resistance from junction to ambient is the sum of the thermal resistance from junction to case (obtained from the transistor's specification sheet) and the effective thermal resistance of the heat sink (obtained from the specifications for the heat sink). In order to protect the transistor, a good heat-sink thermal compound should be used.

Sample Program:

The example is a design based on the economy-line audio power amplifier transistor TIP29. In this design we target a DC power input of 15 watts, thus permitting the resulting amplifier to operate, without difficulty, at an AC power level of 5 watts. A power supply of 60 volts is available. Since this transistor has a maximum collector junction temperature of 150°C, a heat sink will be required at this power level. We elect to use a heat sink with a thermal resistance from case to ambient of 2.5°C per watt*. This must be added to the junction-to-case thermal resistance of the transistor, which in this example is 4.17°C per watt; the result is a thermal resistance, case to am-

---

*Such as the Wakefield Type 641.

bient, of 6.67°C per watt. In order to affect an average design, we elect to input an average value, approximately 45, for both the DC and AC Beta's of the transistor.

Enter Program 20.

Program Listing

```
10 PRINT "POWER AMP DESIGN"
15 PRINT
20 INPUT "ENTER TARGET VCE= ":VC
E
25 INPUT "ENTER TARGET IC= ":IC
30 INPUT "ENTER TARGET ZO= ":ZOO
35 INPUT "ENTER VCC= ":VCC
40 INPUT "ENTER RS= ":RS
45 INPUT "ENTER AMBIENT TEMP. IN
 DEGREES C= ":TA
50 INPUT "ENTER TOTAL THERMAL RE
SISTANCE-JUNCTION TO AMBIENT= ":
RJA
55 INPUT "BETA DC= ":BDC
60 INPUT "BETA AC= ":BAC
65 INPUT "GAMMA= ":G
70 VRE=VCC-VCE
75 RE=VRE/IC
80 IBMAX=IC/BDC
85 IR1=G*IBMAX
90 PRINT "GAMMA= ";G
95 V0=.7
100 R1=(VCC-VRE-V0)/IR1
105 R2=(VRE+V0)/(IR1-IBMAX)
110 RTH=(R1*R2)/(R1+R2)
115 VTH=(VCC*R2)/(R1+R2)
120 ICC=(BDC*(VTH-V0)/(RTH+(BDC+
1)*RE))
125 VCE=(VCC-ICC*RE)
130 VBE=(VCC-IR1*R1-VRE)
135 PD=VCE*ICC
140 TJ=TA+RJA*PD
145 VT=(273+TJ)/11600
150 GM=ICC/VT
155 RPIE=BAC/GM
160 RBS=(RTH*RS)/(RTH+RS)
165 RD=(RPIE+RBS)/(1+BAC)
170 ZO=(RE*RD)/(RE+RD)
175 RLL=(RE*ZO)/(RE+ZO)
180 RZ=(RPIE+(1+BAC)*RLL)
185 ZIN=(RTH*RZ)/(RTH+RZ)
```

```
190 AVI=((1+BAC)*RLL)/RZ
195 AP=((AVI^2)*ZIN/ZO)
200 IF ZO>ZOO THEN 225
205 G=G-1
210 IF G=1 THEN 225
215 PRINT
220 GOTO 85
225 PRINT
230 PRINT "DESIGN GAMMA= ";G
235 PRINT "VCE= ";VCE
240 PRINT "IC= ";ICC
245 PRINT "VBE= ";VBE
250 PRINT "TJ= ";TJ
255 PRINT "DC POWER IN= ";PD
260 PRINT "GM= ";GM
265 PRINT "RPIE= ";RPIE
270 PRINT "R1= ";R1
275 PRINT "R2= ";R2
280 PRINT "RE= ";RE
285 PRINT
290 PRINT "ZO= ";ZO
295 PRINT "ZIN= ";ZIN
300 PRINT "AVI= ";AVI
305 PRINT "AP= ";AP
310 END
```

Run

```
POWER AMP DESIGN

TARGET VCE=  30
TARGET IC=  .5
TARGET ZO=  8


VCC=  60
RS=  1000
AMBIENT TEMP=  25
TOTAL THERMAL RESISTANCE-JUNCTIO
N TO AMBIENT=  6.67
BETA DC=  45
BETA AC=  45
INITIAL GAMMA=  5

DESIGN GAMMA=  2
VCE=  30.49280399
IC=  .5
VBE=  .7
TJ=  125.0230025
```

```
DC POWER IN=  14.9959524
GM=  14.33265044
RPIE=  3.139684471
R1=  1318.5
R2=  2763.
RE=  60

ZO=  8.806046811
ZIN=  254.6862133
AVI=  .9911899288
AP=  28.41438154
```

A few comments are in order. We note that the transistor's junction operating temperature is 125°C. This is within the transistor's specification of 150°C. The input impedance of 255 ohms should make the power amplifier easy to drive. Since the power gain is also high, the circuit will require only about 180 milliwatts to produce an AC power output of 5 watts. This may easily be obtained by means of a voltage amplifier such as that designed by Program 5.

Other Applications:

Program 20 may also be used to design a voltage amplifier. The user need only change program line 95 to:

95 VO = .65

thus providing a better estimator for the base-emitter voltage for a voltage amplifier. The resulting program should be RUN in the same manner as for a power amplifier; however, a hint is in order.

If the initial choices for VCE and IC result in an output impedance that is too small, even with Gamma equal to about 10, the user should try *increasing* VCE and *decreasing* IC. This will result in an increase in the output impedance of the design. An example follows.

Sample Run:

Design an emitter-follower voltage amplifier with an output impedance of about 52 ohms, taking VCC = 9 volts and RS = 6.8K ohms (the output impedance at mid-band of the driving stage). Assume an ambient temperature of 25°C and use a high-Beta transistor, such as the 2N5089.

Run

```
VOLTAGE AMP DESIGN

TARGET VCE=  6
TARGET IC=  .0006
TARGET ZO=  52

VCC=  9
RS=  6800
AMBIENT TEMP=  25
TOTAL THERMAL RESISTANCE-JUNCTIO
N TO AMBIENT=  360
BETA DC=  450
BETA AC=  450
INITIAL GAMMA=  20

DESIGN GAMMA=  20
VCE=  6.006413389
IC=  .0006
VBE=  .65
TJ=  26.29461175
DC POWER IN=  .0035961437
GM=  .0232049648
RPIE=  19392.40173
R1=  200625.
R2=  144078.9474
RE=  5000

ZO=  56.30408675
ZIN=  29073.51273
AVI=  .5642429283
AP=  164.395467
```

A Simplified Mid-Band-, Low-, and High-Frequency Analysis for the Emitter-Follower Amplifier

Circuit Description and Summary of the Analysis

This program is simplified, in that only one value is used for both Beta DC and Beta AC, rather than a "spread". This simplification permits the writing of a program that will do mid-band-, low-, and high-frequency analysis, using a compatible form of the BASIC Language that should run on any machine, and yet be limited to some 80 lines. The user may choose to RUN this program once, using average values for the Beta's, or twice, using the minimum and maximum values. For a complete DC and mid-band analysis without the low- and high-frequency parts, see Program 8.

Program 21 occurs at this point in the book because it can be used to analyze the results of a design from Program 20. In addition to calculating the DC and AC results, the program will also calculate the junction temperature of the transistor, a point of importance in the analysis of Power Amplifiers. In addition to the usual INPUT, the user must specify the base-emitter voltage V-BE. Typical values are 0.65 volts for voltage amplifiers, and 0.7 volts for power amplifiers. If a typical value is specified by the transistor's manufacturer, it may be INPUTed at this point in the program.

The circuit is illustrated in Figure 21.

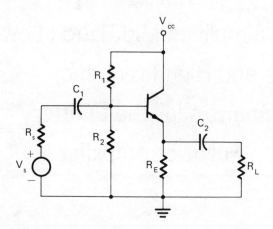

FIGURE 21. A single stage emitter follower amplifier.

As in any high-frequency analysis, the user should refer to the manufacturer's specifications for the transistor's internal capacitances, C-PI and C-MU. Since the emitter follower does not suffer from the Miller Effect, the high frequency 3dB point will exceed that of a simular common-emitter circuit. In view of this, the 3dB point of an actual circuit will be only as "good" as the engineering practice of the user. Care must be taken to reduce all stray capacitance to ground if the actual circuit is to approach the theoretical high-frequency 3dB point. As usual, all parameters in the PRINT statements may be identified in the Glossary at the end of this book.

Sample Programs:

The examples are the results of the two designs obtained previously by means of Program 20. The two designs have been reduced to practice by using the nearest 5% resistors and available capacitors.

The first analysis is for the power amplifier.

Enter Program 21.

Program Listing

```
A SIMPLIFIED MID-BAND, LOW AND
HIGH FREQUENCY ANALYSIS FOR THE
```

```
EMITTER FOLLOWER AMPLIFIER

ALL VOLTAGE IN VOLTS, ALL CURENT
 IN AMPS, ALL RESISTANCE IN OHMS

R1=  1300
R2=  2700
RE=  62
RL=  8
RS=  1000
VCC=  60
BETA DC=  45
BETA AC=  45
AMBIENT TEMP=  25
THERMAL RESISTANCE RJA=  6.67
V-BE=  .7

VCE=  30.22603566
IC=  .4802252313
PD(VCE*IC)=  14.51530497
JUNCTION TEMP=  121.8170841
GM=  14.10935065
RPIE=  3.189374275
ZIN=  239.3550624
Z-OUT=  8.780894922

AVI=  .990309748
AVS=  .1912572585
AI=  29.62945644
AP=  29.34233954

A=  1

LOW FREQ ANALYSIS

C1=  .00001
C2=  .0001

THE LOW FREQ 3 DB POINT IS THE L
ARGER OF THE FOLLOWING TWO FREQ.
FC1=  12.84175519
FC2=  94.84294122


THIS COMPLETES YOUR ANALYSIS
```

Run

```
10 PI=3.141592654
15 PRINT "A SIMPLIFIED MID-BAND,
 LOW AND HIGH FREQUENCY ANALYSIS
```

```
 FOR THE EMITTER FOLLOWER AMPLIF
IER"
20 PRINT
25 PRINT "ALL VOLTAGE IN VOLTS,
ALL CURRENT IN AMPS, ALL RESISTA
NCE IN OHMS"
30 PRINT
35 INPUT "R1= ":R1
40 INPUT "R2= ":R2
45 INPUT "RE= ":RE
50 INPUT "RL= ":RL
55 INPUT "RS= ":RS
60 INPUT "VCC= ":VCC
65 INPUT "BETA DC= ":BDC
70 INPUT "BETA AC= ":BAC
75 INPUT "AMBIENT TEMP. IN DEGRE
ES C = ":TA
80 INPUT "THERMAL RESISTANCE, JU
NCTION TO AMBIENT IN DEGREES C P
ER WATT = ":RJA
85 INPUT "V-BE= ":VO
89 REM  CALCULATE THE DC OPERATI
NG POINT
90 RB=(R1*R2)/(R1+R2)
95 VTH=(VCC*R2)/(R1+R2)
100 IC=((BDC*(VTH-VO))/(RB+(1+BD
C)*RE))
105 VCE=(VCC-(RE*IC))
110 PD=VCE*IC
115 PRINT
120 PRINT "VCE= ";VCE
125 PRINT "IC= ";IC
130 PRINT "PD(VCE*IC)= ";PD
135 TJ=(TA+RJA*PD)
140 PRINT "JUNCTION TEMP= ";TJ
144 REM  CALCULATE THE AC OPERAT
ING POINT AND ALL MID-BAND GAINS

145 VT=(273+TJ)/11600
150 GM=(IC/VT)
155 RPIE=(BAC/GM)
160 PRINT "GM= ";GM
165 PRINT "RPIE= ";RPIE
170 RL1=(RE*RL)/(RE+RL)
175 Z1N=(RPIE+(1+BAC)*RL1)
180 ZIN=(RB*Z1N)/(RB+Z1N)
185 PRINT "ZIN= ";ZIN
190 RSB=(RB*RS)/(RB+RS)
195 ZO1=(RPIE+RSB)/(1+BAC)
200 ZO=(RE*ZO1)/(RE+ZO1)
```

```
205 PRINT "Z-OUT= ";ZO
210 AVI=((1+BAC)*RL1)/Z1N
215 AVS=(AVI*ZIN)/(RS+ZIN)
220 AI=(AVI*ZIN)/RL
225 AP=AI*AVI
230 PRINT
235 PRINT "AVI= ";AVI
240 PRINT "AVS= ";AVS
245 PRINT "AI= ";AI
250 PRINT "AP= ";AP
255 PRINT
260 PRINT "TO DO A LOW FREQ ANAL
YSIS ENTER 1. TO OMIT ENTER 0"
265 INPUT A
270 IF A=1 THEN 275 ELSE 320
275 PRINT
279 REM  CALCULATE THE LOW FREQU
ENCY 3 DB POINT
280 INPUT "C1= ":C1
285 INPUT "C2= ":C2
290 FC1=((1/((RS+ZIN)*C1))/(2*PI
))
295 FC2=((1/((RL+ZO)*C2))/(2*PI)
)
300 PRINT
305 PRINT "THE LOW FREQ 3 DB POI
NT IS THE LARGER OF THE FOLLOWIN
G TWO FREQ."
310 PRINT "FC1= ";FC1
315 PRINT "FC2= ";FC2
320 PRINT
325 PRINT "TO DO A HIGH FREQ ANA
LYSIS INPUT 1. TO OMIT ENTER 0"
330 INPUT B
335 IF B=1 THEN 340 ELSE 400
340 PRINT
344 REM  CALCULATE THE HIGH FREQ
UENCY 3 DB POINT
345 INPUT "C-PIE= ":CP
350 INPUT "C-MU= ":CU
355 GE=((1/RSB)+(1/RPIE)+(RL1/RS
B)*(GM+(1/RPIE)))
360 CE=((1+(RL1/RSB))*CP+(1+(GM*
RL1)+(RL1/RPIE))*CU)
365 CR=((CE^2)-(4*CU*CP*RL1*GE))
370 WHA=(1/(2*CU*CP*RL1))
375 FH=(WHA*(CE-SQR(CR))/(2*PI))
380 PRINT
385 PRINT "HIGH FREQ 3 DB POINT
= ";FH
```

```
390 PRINT
395 PRINT "THIS COMPLETES YOUR A
NALYSIS"
400 END
```

The second analysis is for the voltage amplifier.

Run

```
A SIMPLIFIED MID-BAND, LOW AND
HIGH FREQUENCY ANALYSIS FOR THE
EMITTER FOLLOWER AMPLIFIER

ALL VOLTAGE IN VOLTS, ALL CURENT
 IN AMPS, ALL RESISTANCE IN OHMS

R1=  200000
R2=  150000
RE=  5100
RL=  52
RS=  6800
VCC=  9
BETA DC=  450
BETA AC=  450
AMBIENT TEMP=  25
THERMAL RESISTANCE RJA=  360
V-BE=  .65

VCE=  5.914934703
IC=  .0006049148
PD(VCE*IC)=  .0035780313
JUNCTION TEMP=  26.28809128
GM=  .0234456748
RPIE=  19193.30553
ZIN=  28371.37853
Z-OUT=  55.90693461

AVI=  .5474195027
AVS=  .4415819503
AI=  298.6739602
AP=  163.4999508

A=  1

LOW FREQ ANALYSIS

C1=  .000003
C2=  .000015
```

```
THE LOW FREQ 3 DB POINT IS THE L
ARGER OF THE FOLLOWING TWO FREQ.
FC1=  1.508375557
FC2=  98.32852333

B=  1

HIGH FREQ ANALYSIS

C-PIE=  1.6E-11
C-MU=  2.E-12

HIGH FREQ 3 DB POINT=
 3124640.694

THIS COMPLETES YOUR ANALYSIS
```

Note: These circuits are good 'benchmarks' that the user can modify to suit his or her needs. Suggestions for modification by means of design are included in the description of Program 20.

Design of a FET Tuned-Drain Oscillator

Circuit Description and Summary of the Analysis

The tuned-drain field-effect transistor oscillator is a simple yet effective generator of a sinusoidal signal at high frequencies. It may easily be used to frequencies as high as the transistor's bandwidth. The circuit is illustrated in Figure 22.

The design algorithm requires the user to first pick the desired frequency of oscillation, f_o. A convenient value for the capacitor C must then be specified, on the basis that the frequency of oscillation is approximately related to C and L_1 by the equation:

$$f_o \cong \frac{1}{2\pi \sqrt{L_1 \cdot C}} \qquad (1)$$

The program calculates the required inductance L_1. (For simplicity of construction the coils L_1 and L_2 may be wound on a common low-loss core so as to make the mutual inductance, M, easily adjustable.) In general M is bounded by the condition:

$$M \leq \sqrt{L_1 L_2} \qquad (2)$$

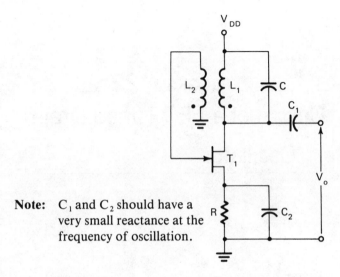

Note: C_1 and C_2 should have a very small reactance at the frequency of oscillation.

FIGURE 22. A tuned-drain field effect transistor oscillator.

Because the gate of the FET draws very little current, the inductance L_2 is not critical; however, M must be large enough to insure oscillation, given the maximum g_m of the particular transistor used in the circuit. A simple choice is to make M from 3% to 5% of L_1. This is easily reduced to practice.

Once the mutual inductance is fixed by the user, the program will calculate the minimum value of g_m required for oscillation. The user must next INPUT typical values for the transistor's IDSS and VP. The program then calculates the absolute value of the maximum gate-source voltage that may be used in the circuit. The user must specify a value for VGS. This determines the value of the resistor R that will be required to bias the transistor.

The program then calculates the minimum value needed for the power supply voltage VDD. The user should choose and INPUT a voltage that exceeds this, by a factor of at least two for good operation. The program then calculates and PRINTS all the needed information for the actual construction of the tuned-drain oscillator.

If the oscillator fails to oscillate when constructed, try increasing M by moving coil L_2 closer to coil L_1. Construction data for the coils may be found in either of the references below.

1. Electronics Engineers' Handbook, Edited by Donald G. Fink and Donald Christiansen, McGraw-Hill Book Company, 1982.

2. The Radio Amateur's Handbook, American Radio Relay League, Newington, Conn., USA 1982.

Sample Program:

The sample program designs an oscillator with a frequency of 10 mHz using a capacitor of 50 pF. A coil Q of 100 is assumed as this is easily obtained in practice. As our example we have elected to use transistor type 2N5359. From the specification sheet, note that the absolute value of the small-signal common-source output admittance is at most 10 micro-mhos; therefore, take RD = 100K ohms. As INPUT for the mutual inductance, take M equal to 4% of L_1 or 0.2 microhenry. From the specification sheet, note that the average values for IDSS and VP are:

$$IDSS = 1.2mA; \qquad VP = -1.2\,V.$$

The maximum transconductance, g_m, may be calculated by means of the equation

$$\text{Max } g_m = -\frac{2 * IDSS}{VP}.$$

It is 2.0E-3 for the average characteristics. The calculated minimum value required for oscillation is 1.049E-3; hence the design will be feasible with this transistor. Next INPUT the average values for IDSS and VP, taking care to include the correct sign for the pinch voltage (negative for the N-JFET). The computer should respond that the maximum absolute value of the gate-source voltage, VGS, is about 0.57 volts. Next INPUT a value of − .4 volts. The computer should respond that the minimum power supply voltage required to insure operation of the transistor in its saturation region is 1.8 volts. A power supply voltage of 6 volts will be more than adaquate. This concludes the design.

Enter Program 22.

Program Listing

```
10 PI=3.141592654
15 PRINT "FET TUNED DRAIN OSCILL
ATOR DESIGN"
20 PRINT
24 REM   DO THE AC DESIGN
25 INPUT "OSC. FREQ. IN HZ= ":F
30 INPUT "CAPACITOR C= ":C
35 L=(1/(((2*PI*F)^2)*C))
40 PRINT "L1= ";L
45 INPUT "Q OF COIL L1= ":Q
50 RS=((2*PI*F*L)/Q)
55 INPUT "RD= ":RD
```

```
60 INPUT "MUTUAL INDUCTANCE M= "
:M
65 REFF=(Q*Q*RS)
70 PRINT "R EFF= ";REFF
75 RDR=(RD*REFF)/(RD+REFF)
80 GMMIN=(L/(M*RDR))
85 PRINT
90 PRINT "GM MIN FOR OSCILATION=
";GMMIN
95 PRINT
99 REM  DO THE DC DESIGN
100 INPUT "IDSS= ":IDSS
105 PRINT
110 PRINT "THE USER MUST ENTER T
HE PINCH OFF VOLTAGE VP. NOTE VP
 IS NEGATIVE FOR AN NJFET"
115 INPUT "VP= ":VP
120 VGSMAX=((((ABS(VP))^2)/(2*ID
SS))*(((2*IDSS)/(ABS(VP)))-GMMIN
))
125 PRINT
130 PRINT "ABS(VGS MAX)= ";VGSMA
X
135 PRINT
140 PRINT "THE USER MUST SPECIFY
 VGS. FOR A NJFET,VGS IS A NEGAT
IVE VOLTAGE."
145 PRINT "PICK A VALUE BETWEEN
VGS MAX AND ZERO"
150 INPUT "VGS= ":VGS
155 ID=(IDSS*((1-(VGS/VP))^2))
160 R=(ABS(VGS)/ID)
165 VDDMIN=(VGS-VP)
170 PRINT "THE MINIMUM POWER SUP
PY VOLTAGE VDD= ";(VDDMIN+1)
175 PRINT
180 INPUT "SPECIFY THE POWER SUP
PLY VOLTAGE VDD= ":VDD
185 VDS=(VDD-(ID*R))
190 PRINT
194 REM  PRINT THE RESULTS
195 PRINT "VDD= ";VDD
200 PRINT "VDS= ";VDS
205 PRINT "ID= ";ID
210 PRINT "R= ";R
215 PRINT "L1= ";L
220 PRINT "M= ";M
225 PRINT "C= ";C
230 PRINT "F= ";F
235 END
```

Run

```
FET TUNED DRAIN OSCILLATOR
 DESIGN

OCS. FREQ. IN HZ=  10000000
CAPACITOR C=  5.E-11
L1=  5.06606E-06
Q OF COIL L1=  100

RD=  100000
MUTUAL INDUCTANCE M=  .0000002

GM MIN FOR OSCILATION=
 .0010490777

IDSS=  .0012
VP= -1.2

ABS(VGS MAX)=  .5705533954

VGS= -.4
THE MINIMUM POWER SUPPLY VOLTAGE
 VDD=  1.8


VDD=  6
VDS=  5.6
ID=  .0005333333
R=  750
L1=  5.06606E-06
M=  .0000002
C=  5.E-11
F=  10000000
```

The Transfer Characteristics of an Integrated CMOS Inverter

Circuit Description and Summary of the Analysis

This program has been included in order to illustrate the simulation of a simple two-transistor integrated circuit on the micro-computer. The CMOS Inverter, or NOT gate, is a good choice for such a simulation because the transfer characteristics may be modeled by way of a few non-linear equations. This non-linear description is easily simulated by a program of some 75 lines which should be understandable at the undergraduate level. The CMOS NOT gate is illustrated in Figure 23.

The model for the MOSFET transistors is based on that found in a standard text in digital electronics.[1] While other, more sophisticated models are possible, the reference to a standard text is of value to the user. The analytic expressions for the current in an n-channel MOSFET are given by:

$$I_{DS} = k\,[2(V_{GS} - V_T)\,V_{DS} - V_{DS}^2]\,, \text{ for } 0 \le V_{DS} \le V_{GS} - V_T \qquad (1)$$

The Ohmic Region

$$I_{DS} = k\,(V_{GS} - V_T)^2\,, \text{ for } 0 \le V_{GS} - V_T \le V_{DS} \qquad (2)$$

The Saturation Region

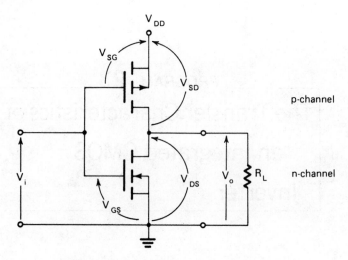

FIGURE 23a. An integrated circuit CMOS inverter.

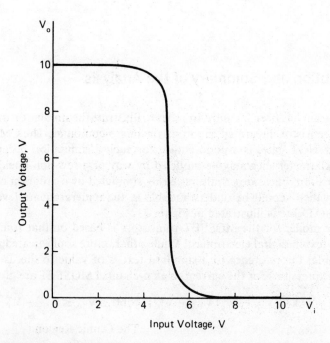

FIGURE 23b. Transfer Characteristics for a Type CD4007A Inverter as Simulated by Program 23.

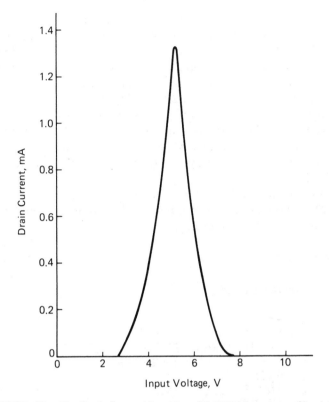

FIGURE 23c. The Drain Current for a Type CD4007A Inverter as Simulated by Program 23.

Values for the threshold voltage V_T and the constant k may be found in the data supplied with the specifications for any particular gate. The equations for the p-channel device may be obtained from the above by simply reversing all subscripts; for example, "I_{DS}" becomes "I_{SD}".

The program was written so as to take advantage of the fact that there are four regions of operation, during each of which various combinations of the above equations must be used for the n- and p-channel devices. These regions have been labeled in the program and in the print statements, for further study by the user.

Sample Program:

The sample program simulates the type 4007 CMOS NOT gate, such as the RCA CD4007A[2]. From the specifications, note that the threshold voltage V_T for the n- and p-channel devices is about 2.5 volts. This may be obtained from a study of Figure 2 of the RCA reference. The values for the constants

k_n and k_p may be obtained from Figures 8 and 6 of the same reference, together with equation (2). For example, for the case of a typical n-channel MOSFET with VGS = 10 volts, note from Figure 8 that the drain current is 12 mA in the saturation region. Equation (2) then yields the value 1.875×10^{-4} amps/volts^2 for k_n. A similar procedure for the typical p-channel device yields a value of 2.5×10^{-4} for k_p.

Now elect to simulate the operation of an unloaded gate with a power-supply voltage of 10 volts. Since "infinity" is not an allowable input for R_L, take $R_L = 1 \times 10^{12}$ ohms, a value which is sufficiently large so as to simulate the no-load result. The user may wish to try this program for other values of VDD and RL.

This program also prints the value for the drain current I_D as defined in Figure 23. The sample includes the results of this simulation and plots of the transfer characteristics and of the current I_D.

Enter Program 23.

Program Listing

```
10 P=0
15 DIM I(200)
20 PRINT "TRANSFER CHARACTERISTI
C OF AN IC CMOS INVERTER"
25 PRINT
30 INPUT "VSS= ":VSS
35 PRINT
40 INPUT "VT-N= ":VTN
45 PRINT
50 INPUT "VT-P= ":VTP
55 PRINT
60 INPUT "KN= ":KN
65 PRINT
70 INPUT "KP= ":KP
75 PRINT
80 INPUT "RL= ":RL
85 PRINT
90 INPUT "VI(MAX,MIN)":VIMAX,VIM
IN
95 PRINT
100 INPUT "NO OF STEPS= ":M
105 PRINT
110 VO=VSS
115 H=(VIMAX-VIMIN)/M
```

1. Herbert Taub and Donald Schilling, "Digital Integrated Electronics", McGraw-Hill Book Co., 1977.

2. "RCA Integrated Circuits", The Radio Corporation of America, 1976.

```
120 PRINT
125 FOR N=0 TO M STEP 1
130 VI=VIMIN+N*H
135 PRINT "VI= ";VI
140 IF VI<VTN THEN 150
145 IF VI>VTN THEN 200
149 REM   DO FOR REGION #1
150 PRINT "R(#1)"
155 A1=((1/(KP*RL))-2*(VI+VTP))
160 B1=(2*(VI+VTP)*VSS-VSS^2)
165 Y=SQR(A1^2-4*B1)
170 VO=(-A1+Y)/2
175 PRINT "VO= ";INT(VO*1E3)/1E3
180 I(N)=KP*(2*(VSS-VI-VTP)*(VSS
-VO)-((VSS-VO)^2))
185 PRINT "I(";N;")=";INT(I(N)*1
E6)/1E6
190 PRINT
195 IF VI<=VTN THEN 360
199 REM    DO FOR REGION #2
200 IF VI>(VO-VTP)THEN 265
205 PRINT "R(#2)"
210 A2=((1/(KP*RL))-2*(VI+VTP))
215 B2=(2*(VI+VTP)*VSS-VSS^2+(KN
/KP)*((VI-VTN)^2))
220 X2=(A2^2-4*B2)
225 IF X2<0 THEN 265
230 Y2=SQR(X2)
235 VO=(-A2+Y2)/2
240 PRINT "VO= ";INT(VO*1E3)/1E3
245 I(N)=KP*(2*(VSS-VI-VTP)*(VSS
-VO)-((VSS-VO)^2))
250 PRINT "I(";N;")=";INT(I(N)*1
E6)/1E6
255 PRINT
260 IF VI<(VO-VTP)THEN 360
264 REM  DO FOR REGION #3
265 IF VI>VO+VTN THEN 305
270 PRINT "R(#3)"
275 VO=(RL*KP*((VSS-VI-VTP)^2)-K
N*RL*((VI-VTN)^2))
280 PRINT "VO= ";INT(VO*1E3)/1E3
285 I(N)=KP*((VSS-VI-VTP)^2)
290 PRINT "I(";N;")=";INT(I(N)*1
E6)/1E6
295 PRINT
300 IF VI<(VO+VTN)THEN 360
304 REM  DO FOR REGION #4
305 PRINT "R(#4)"
310 IF VI>VSS-VTP THEN 370
```

```
315 A4=-((1/(KN*RL))+2*(VI-VTN))
320 B4=(KP/KN)*((VSS-VI-VTP)^2)
325 X4=(A4^2-4*B4)
330 Y4=SQR(X4)
335 VO=(-A4-Y4)/2
340 PRINT "VO= ";INT(VO*1E3)/1E3
345 I(N)=KP*((VSS-VI-VTP)^2)
350 PRINT "I(";N;")=";INT(I(N)*1
E6)/1E6
355 PRINT
360 P=P+VSS*I(N)
365 NEXT N
370 AP=P/N
375 PRINT
380 PRINT "AVERAGE POWER= ";AP
385 END
```

Run

```
TRANSFER CHARACTERISTIC OF AN IC
 CMOS INVERETR

VSS= 10
VT-N= 2.5
VT-P= 2.5
KN= .0001875
KP= .00025
RL= 1.E+12
VIMAX= 10 VIMIN= 0
NO OF STEPS= 50

VI= 0
R(#1)
VO= 9.999
I( 0 )= 0

VI= .2
R(#1)
VO= 9.999
I( 1 )= 0

VI= .4
R(#1)
VO= 9.999
I( 2 )= 0

VI= .6
R(#1)
VO= 9.999
I( 3 )= 0
```

```
VI=  .8
R(#1)
VO=  9.999
I( 4 )= 0

VI=  1
R(#1)
VO=  9.999
I( 5 )= 0

VI=  1.2
R(#1)
VO=  9.999
I( 6 )= 0

VI=  1.4
R(#1)
VO=  9.999
I( 7 )= 0

VI=  1.6
R(#1)
VO=  9.999
I( 8 )= 0

VI=  1.8
R(#1)
VO=  9.999
I( 9 )= 0

VI=  2
R(#1)
VO=  9.999
I( 10 )= 0

VI=  2.2
R(#1)
VO=  9.999
I( 11 )= 0

VI=  2.4
R(#1)
VO=  9.999
I( 12 )= 0

VI=  2.6
R(#2)
VO=  9.999
I( 13 )= .000001
```

```
VI=  2.8
R(#2)
VO=  9.992
I( 14 )= .000016

VI=  3
R(#2)
VO=  9.979
I( 15 )= .000046

VI=  3.2
R(#2)
VO=  9.957
I( 16 )= .000091

VI=  3.4
R(#2)
VO=  9.925
I( 17 )= .000151

VI=  3.6
R(#2)
VO=  9.881
I( 18 )= .000226
VI=  3.8
R(#2)
VO=  9.824
I( 19 )= .000316

VI=  4
R(#2)
VO=  9.749
I( 20 )= .000421

VI=  4.2
R(#2)
VO=  9.653
I( 21 )= .000541

VI=  4.4
R(#2)
VO=  9.527
I( 22 )= .000676

VI=  4.6
R(#2)
VO=  9.358
I( 23 )= .000826

VI=  4.8
```

```
R(#2)
VO=   9.122
I( 24 )= .000991

VI=   5
R(#2)
VO=   8.749
I( 25 )= .001171

VI=   5.2
R(#2)
R(#3)
VO= -44375000
I( 26 )= .001322
```
⎫
⎬ Overshoot
⎭

```
R(#4)
VO=   2.213
I( 26 )= .001322

VI=   5.4
R(#4)
VO=   1.309
I( 27 )= .001102

VI=   5.6
R(#4)
VO=   .909
I( 28 )= .000902
VI=   5.8
R(#4)
VO=   .647
I( 29 )= .000722

VI=   6
R(#4)
VO=   .458
I( 30 )= .000562

VI=   6.2
R(#4)
VO=   .318
I( 31 )= .000422

VI=   6.4
R(#4)
VO=   .212
I( 32 )= .000302

VI=   6.6
R(#4)
```

```
VO=  .133
I( 33 )= .000202

VI=  6.8
R(#4)
VO=  .076
I( 34 )= .000122

VI=  7
R(#4)
VO=  .037
I( 35 )= .000062

VI=  7.2
R(#4)
VO=  .012
I( 36 )= .000022

VI=  7.4
R(#4)
VO=  .001
I( 37 )= .000002

VI=  7.6
R(#4)

AVERAGE POWER= .0029564145
```

Design of an Unregulated Full-Wave Bridge Rectifier Power Supply

Circuit Description and Summary of the Analysis

The availability of inexpensive integrated-circuit regulator chips has simplified the design of a complete regulated power supply. Such a regulated supply contains two principal parts:

1. The unregulated power supply with its filter.
2. The integrated circuit regulator.

This program may be used to design the unregulated power supply and its filter. The user may elect to design either a capacitor-filtered or a pi-filtered full-wave bridge rectifier. The analysis for the power supply is readily available in the literature; the reader should consult the reference below for complete details.

Reference: Michael M. Cirovic, ''Basic Electronics: Devices, Circuits and Systems'', Reston Publishing Co. Second Edition, 1979.

The full-wave bridge rectifier is illustrated in Figure 24. In the design algorithm, the two capacitors of the Pi Filter have been taken equal.

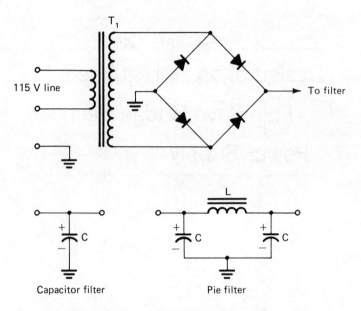

Capacitor filter   Pie filter

FIGURE 24. An unregulated full-wave bridge rectifier.

The design procedure requires the user to INPUT the desired DC load voltage and DC load current, as well as the desired ripple factor, r, defined by the equation:

$$r = \frac{\text{RMS Ripple Voltage at the Output}}{\text{DC Load Voltage}} \tag{1}$$

When a pi-filter design is contemplated, the user must also specify the inductance of the filter choke and its DC resistance.

The program will calculate the value of the required filter capacitor, the percentage of regulation, the peak transformer voltage, the RMS transformer voltage on the secondary winding, the peak-to-peak ripple voltage, the peak diode forward current, and the peak reverse diode voltage. While the percentage of regulation, as defined by the equation:

$$\% \text{ Regulation} = \frac{\text{No Load Voltage} - \text{Full Load Voltage}}{\text{Full Load Voltage}} \times 100,$$

may be large for the unregulated power-supply design, it should be noted that unregulated power supplies may suffice in applications where the load current does not vary a great deal.

Sample Programs:

The sample programs illustrate the design of two power supplies, one using a capacitor filter and the other using a pi filter. Both power supplies are designed to deliver a full-load voltage of 12 volts and a full-load current of 1 ampere. The user should note that while the percentage of regulation of the power supply using the pi filter is greater than that of the power supply using the capacitor filter, the peak-to-peak ripple voltage is much less. The regulation of the pi-filter supply may be improved by decreasing the designed ripple factor.

Enter Program 24.

Program Listing

```
10 PRINT "DESIGN OF AN UNREGULAT
ED FULL WAVE BRIDGE RECTIFIER"
15 PRINT
20 PRINT "ALL VOLTAGES IN VOLTS.
 ALL CURRENTS IN AMPS."
25 PRINT
30 INPUT "TARGET DC OUTPUT VOLTA
GE IN VOLTS= ":VDC
35 INPUT "TARGET DC OUTPUT CURRE
NT IN AMPS= ":IDC
40 INPUT "TARGET RIPPLE FACTOR=
":RP
45 INPUT "EXPECTED DIODE VOLTAGE
 DROP IN VOLTS= ":V0
50 PRINT
55 PRINT "DESIGNER MAY CHOOSE A
CAPACITOR FILTER OR A PIE FILTER
"
60 PRINT "ENTER 1 FOR THE CAPACI
TOR FILTER DESIGN"
65 PRINT "ENTER 2 FOR THE PIE FI
LTER DESIGN"
7u INPUT A
75 IF A=1 THEN 95
80 IF A=2 THEN 220
85 PRINT "TRY AGAIN"
90 GOTO 55
95 RL=(VDC/IDC)
100 C=(2400/(RL*RP))
105 VM=(VDC+((4200*IDC)/C)+(2*V0
))
```

```
110 REG=((VM-VDC)/VDC)*100
115 TSV=(VM/(SQR(2)))
120 PPRV=(2*RP*VDC*(SQR(3)))
125 PDFC=((10*VM)/RL)
130 PRDV=2*VM
135 PRINT
140 PRINT "CAPACITOR FILTER DESI
GN"
145 PRINT
150 PRINT "REQUIRED FILTER CAPAC
ITOR IN MICROFARADS= ";C
155 PRINT
160 PRINT "% REGULATION= ";REG
165 PRINT
170 PRINT "PEAK TRANSFORMER VOLT
AGE= ";VM
175 PRINT
180 PRINT "SECONDARY RMS TRANSFO
RMER VOLTAGE= ";TSV
185 PRINT
190 PRINT "PEAK-TO-PEAK RIPPLE V
OLTAGE= ";PPRV
195 PRINT
200 PRINT "PEAK DIODE FORWARD CU
RRENT= ";PDFC
205 PRINT
210 PRINT "PEAK REVERSE DIODE VO
LTAGE= ";PRDV
215 END
220 RL=(VDC/IDC)
225 PRINT
230 PRINT "PIE FILTER DESIGN"
235 PRINT
240 PRINT "CHOOSE A CHOKE"
245 INPUT "INDUCTANCE OF CHOKE I
N HENRIES= ":L
250 INPUT "DC RESISTANCE OF CHOK
E IN OHMS= ":RC
255 C=SQR(3300/(RP*RL*L))
260 VM=(VDC+((4200*IDC)/C)+(2*V0
)+(IDC*RC))
265 REG=((VM-VDC)/VDC)*100
270 TSV=(VM/(SQR(2)))
275 PPRV=(SQR(2)*RP*VDC)
280 PDFC=((10*VM)/RL)
285 PRDV=2*VM
290 PRINT
295 PRINT "REQUIRED FILTER CAPAC
ITORS IN MICROFARADS= ";C
300 PRINT
```

```
305 PRINT "% REGULATION= ";REG
310 PRINT
315 PRINT "PEAK TRANSFORMER VOLT
AGE= ";VM
320 PRINT
325 PRINT "SECONDARY RMS TRANSFO
RMER VOLTAGE= ";TSV
330 PRINT
335 PRINT "PEAK-TO-PEAK RIPPLE V
OLTAGE= ";PPRV
340 PRINT
345 PRINT "PEAK DIODE FORWARD CU
RRENT= ";PDFC
350 PRINT
355 PRINT "PEAK REVERSE DIODE VO
LTAGE= ";PRDV
360 END
```

Run

```
DESIGN OF AN UNREGULATED FULL
 WAVE BRIDGE RECTIFIER

ALL VOLTAGES IN VOLTS.
ALL CURRENTS IN AMPS.

TARGET DC OUTPUT VOLTAGE=  12
TARGET DC OUTPUT CURRENT=  1
TARGET RIPPLE FACTOR=  .003
DIODE VOLTAGE DROP=  1

CAPACITOR FILTER DESIGN

REQUIRED FILTER CAPACITOR IN MIC
ROFARADS=  66666.66667

% REGULATION=  17.19166667

PEAK TRANSFORMER VOLTAGE=
 14.063

SECONDARY RMS TRANSFORMER VOLTAG
E=  9.944042664

PEAK-TO-PEAK RIPPLE VOLTAGE=
 .1247076581

PEAK DIODE FORWARD CURRENT=
 11.71916667
```

```
PEAK REVERSE DIODE VOLTAGE=
 28.126

DESIGN OF AN UNREGULATED FULL
 WAVE BRIDGE RECTIFIER

ALL VOLTAGES IN VOLTS.
ALL CURRENTS IN AMPS.

TARGET DC OUTPUT VOLTAGE=  12
TARGET DC OUTPUT CURRENT=  1
TARGET RIPPLE FACTOR=  .00001
DIODE VOLTAGE DROP=  1

PIE FILTER DESIGN

CHOKE INDUCTANCE=  10
CHOKE DC RESISTANCE=  .75

REQUIRED FILTER CAPACITORS IN MI
CROFARADS=  1658.312395

% REGULATION=  44.02246079

PEAK TRANSFORMER VOLTAGE=
 17.28269529

SECONDARY RMS TRANSFORMER VOLTAG
E=  12.22071104

PEAK-TO-PEAK RIPPLE VOLTAGE=
 .0001697056

PEAK DIODE FORWARD CURRENT=
 14.40224608

PEAK REVERSE DIODE VOLTAGE=
 34.56539059
```

Note: Unregulated power supplies designed for use with a regulator should meet the input voltage requirement of the regulator, which depends on the required output voltage and current from the regulator.

PROGRAM **25**

Analysis and Design of
Cascade Multi-Stage
Transistor Amplifiers

Circuit Description and Summary of the Analysis

In view of the uncountable number of possible multi-stage circuits, it is not possible to consider all cases in detail in this work. However, there is a method of approach that uses the preceding programs and may be applied to any circuit containing no more than about four stages. The limit of four stages is dictated by the time involved rather than by the limitations of the programs.

Consider the ith and $(i + 1)$th stages, as illustrated in Figure 25a.

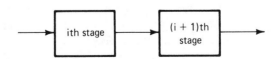

FIGURE 25a. Two typical stages of a multi-stage amplifier.

The load impedance of the ith stage is the input impedance of the (i + 1)th stage, while the source impedance of the (i + 1)th stage is the output impedance of the ith stage. Using these identifications of the load and source impedances, it is possible to calculate the overall mid-band voltage gain of the cascade by simply using the proper program for each stage and then multiplying the individual results by way of the equation:

$$A_{VS} \text{(Cascade)} = A_{VS_1} * A_{Vi_2} * A_{Vi_3} * .... A_{Vi_n} \tag{1}$$

The user should note that while A_{VS}, the overall voltage gain, must be used for the first stage, the following stages require that the circuit voltage gain be used, namely A_{VI}.

An estimate for the low-frequency 3dB point of the cascade may be obtained by summing the low-frequency 3dB points of all the capacitors in the circuit.

$$f_L \text{(cascade)} = \sum_{m=1}^{M} fc_m \tag{2}$$

The low-frequency 3dB points due to all capacitors are added like resistors in series. This is a conservative estimate. In other words, the circuit will almost always be somewhat better than the estimate.

The high-frequency 3dB point for the cascade may be approximated by the reciprocal of the sum of the reciprocals of the high-frequency 3dB points due to all of the transistor's internal capacitances and to the stray capacitances to ground.

$$\frac{1}{f_H \text{(Cascade)}} = \sum_{n=1}^{N} \frac{1}{f_n} \tag{3}$$

The high-frequency 3dB points are added like resistors in parallel. The accuracy of this method will depend on the user's knowledge of the stray capacitances to ground.

The analysis of a cascade amplifier is straightforward, but the design of such a circuit to the user's specifications is not at all direct or unique. For this task some prior knowledge of design is required. While design of a cascade is indeed possible using the design programs in this book, the task is best left to the qualified and interested reader. The example provided is one of analysis. (See Problem 27 for an example of 3-stage design.)

Sample Program:

The example analyzes a two-stage BJT cascade common-emitter amplifier for its overall mid-band voltage gain and its low- and high-frequency 3dB points. The circuit is illustrated in Figure 25b; it is based on a design in reference 1, with a number of changes.

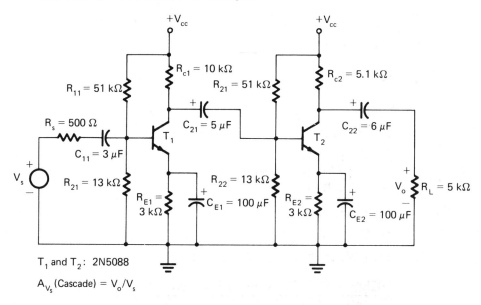

T_1 and T_2: 2N5088

A_{V_s} (Cascade) $= V_o/V_s$

FIGURE 25b. The two stage amplifier used for the sample.

The Second Stage

Begin with an analysis of the second stage, using Program 9. The user should refer to the specifications for a 2N5088 BJT transistor at the end of this book. Since the maximum value of I_c must be less then 2.2 mA in both stages, use the Beta DC and AC specifications for a collector current of 1 mA, namely: $350 \leq$ Beta DC ≤ 1050, $350 \leq$ Beta AC ≤ 1400. For the transistor capacitances, use the typical values of 4 pF and 1.8 pF. This is all the information required; Program 9 yields the following results:

Run

```
COMPLETE ANALYSIS OF A COMMON EM
ITTER AMPLIFIER
R1=   51000
R2=   13000
```

```
RE=  3000

RC=  5100

RL=  5000

RS=  10000

C1=  .000005

C2=  .000006

CE=  .0001

C PIE=  4.E-12

C MU=  1.8E-12

VCC=  18
BETA DC MAX=  1050
BETA DC MIN=  350

BETA AC MAX=  1400
BETA AC MIN=  350

AMBIENT TEMP. IN DEG C = 25

THERMAL RESISTANCE, JUNCTION TO
AMBIENT, IN DEG. C/WATT =  357

I-CEO=  0

ICMAX=  .0009978514
ICMIN=  .0009894938

VCE MAX=  9.993557498
VCE MIN=  9.920252159

T-J MAX DEG C = 28.53021703
T-J MIN DEG C = 28.5339206

ZIN MAX=  8067.719557
ZIN MIN=  4871.150036

AVS MAX=  43.27721128
AVS MIN=  31.4804293

AVI MAX=  96.91964515
AVI MIN=  96.10670661
```

```
MAX CURRENT GAIN=   156.3841033
MIN CURRENT GAIN=   93.63003747

MAX POWER GAIN=   15156.6918
MIN POWER GAIN=   8998.474541

OUTPUT Z=   5100

F C1 MAX=   2.140452493
F C1 MIN=   1.761760253

F C2=   2.626319321

F CE MAX=   54.18445423
F CE MIN=   40.28382716

MAX H.F. 3 D.B. POINT=
  269667.0421
MIN H.F. 3 D.B. POINT=
  196625.2538
```

The specification sheets for the 2N5088 contain a plot of the gain-bandwidth product f_T as a function of the collector DC current; the estimates for the 3dB high-frequency points could be improved by using the equation:

$$C\pi = \frac{g_m}{2\pi f_T} - C_\mu, \tag{4}$$

where the transconductance at the maximum and minimum points for the Beta's may be obtained by printing lines 280 and 285 of the program. However, such detailed data is usually available for a discrete transistor only if measured by the user. In general, f_T is dependent on I_c (DC) and C_μ is dependent on V_{CB} (DC).

The First Stage

The analysis of the first stage again uses the 1 mA data for the Beta's and the typical values for C-PI and C-MU. In addition it requires an estimate of the load impedance, on the first stage, resulting from the presence of the second stage. From the results for the analysis of the second stage, note that Z_{in} varies from a high of 8,068 ohms to a low of 4,871 ohms. As this range is large, perform the analysis for the first stage twice, using the particular value appropriate to each Beta extreme. First, for the case of the lower Beta extreme, take $R_L = 4,871$ ohms:

Run

```
COMPLETE ANALYSIS OF A COMMON EM
ITTER AMPLIFIER

R1=   51000

R2=   13000

RE=   3000

RC=   10000

RL=   4871

RS=   500

C1=   .000003

C2=   .000005

CE=   .0001

C PIE=   4.E-12

C MU=   1.8E-12

VCC=   18
BETA DC MAX=   350
BETA DC MIN=   350

BETA AC MAX=   350
BETA AC MIN=   350

AMBIENT TEMP. IN DEG C = 25

THERMAL RESISTANCE, JUNCTION TO
AMBIENT, IN DEG. C/WATT =   357

I-CEB=   0

ICMAX=   .0009894938
ICMIN=   .0009894938

VCE MAX=   5.145037919
VCE MIN=   5.145037919

T-J MAX DEG C = 26.81748096
T-J MIN DEG C = 26.81748096
```

```
ZIN MAX=   4856.420596
ZIN MIN=   4856.420596

AVS MAX=   113.6929902
AVS MIN=   113.6929902

AVI MAX=   125.398421
AVI MIN=   125.398421


MAX CURRENT GAIN=   125.0230906
MIN CURRENT GAIN=   125.0230906

MAX POWER GAIN=   15677.69815
MIN POWER GAIN=   15677.69815

OUTPUT Z=   10000

F C1 MAX=   9.904310041
F C1 MIN=   9.904310041

F C2=   2.140474089

F CE MAX=   58.60511744
F CE MIN=   59.55293183

MAX H.F. 3 D.B. POINT=
 1435782.955
MIN H.F. 3 D.B. POINT=
 1435782.955
```

Second, for the case of the upper Beta extreme, take R_L = 8,068 ohms:

Run

```
COMPLETE ANALYSIS OF A COMMON EM
ITTER AMPLIFIER

R1=   51000

R2=   13000

RE=   3000

RC=   10000

RL=   8068

RS=   500
```

```
C1=  .000003

C2=  .000005

CE=  .0001

C PIE=  4.E-12

C MU=  1.8E-12

VCC=  18
BETA DC MAX=  1050
BETA DC MIN=  1050

BETA AC MAX=  1400
BETA AC MIN=  1400

AMBIENT TEMP. IN DEG C = 25

THERMAL RESISTANCE, JUNCTION TO
AMBIENT, IN DEG. C/WATT =  357

I-CEO=  0

ICMAX=  .0009978514
ICMIN=  .0009978514

VCE MAX=  5.030780424
VCE MIN=  5.030780424

T-J MAX DEG C = 26.75698991
T-J MIN DEG C = 26.75698991

ZIN MAX=  8057.175842
ZIN MIN=  8057.175842

AVS MAX=  162.3538991
AVS MIN=  162.3538991

AVI MAX=  172.4290112
AVI MIN=  172.4290112


MAX CURRENT GAIN=  172.1976777
MIN CURRENT GAIN=  172.1976777

MAX POWER GAIN=  29691.87529
MIN POWER GAIN=  29691.87529

OUTPUT Z=  10000

F C1 MAX=  6.199668123
F C1 MIN=  6.199668123
```

```
F C2=   1.761732908

F CE MAX=   61.2332774
F CE MIN=   62.22359675

MAX H.F.  3 D.B.  POINT=
  1014457.915
MIN H.F.  3 D.B.  POINT=
  1014457.915
```

Estimate of the Bounds on the Overall Mid-Band Voltage Gain, $A_{v.s}$

The overall mid-band voltage gain may be calculated by using equation (1). For the case of a two-stage amplifier,

$$A_{VS} = A_{VS_1} * A_{Vi_2}.$$

For the upper bound,

$$A_{Vs_{max}} = 162.35 * 96.92 = 15,735 \quad \text{(maximum Beta's)}.$$

For the lower bound,

$$A_{Vs_{min}} = 113.69 * 96.1 = 10,926 \quad \text{(minimum Beta's)}.$$

The overall mid-band voltage gain of the two-stage amplifier should lie somewhere between these two extremes, depending on the actual Beta's of the transistors used for any particular realization. Note that the average mid-band voltage gain for the population is about 13,330. If the user is interested only in obtaining a value for the mid-band gain, some computer time could be saved by running the analysis for the first stage only once, INPUTing the average value for Z_{in_2}, namely 6,470 ohms, as R_L. This procedure results in an average gain for the two-stage voltage amplifier of 13,470, a rather small difference from the more complete result.

Estimate of the Low-Frequency 3dB Points

The low-frequency 3dB points of a cascade may be estimated by using equation (2). For the case of our two-stage amplifier, one must consider the effects of all five capacitors.

For the upper bound (maximum Betas), one obtains:

$$f_L = \underset{C_{11}}{6.2} + \underset{C_{21}}{1.76} + \underset{C_{E_1}}{61.2} + \underset{C_{22}}{2.6} + \underset{C_{E_2}}{54.2} = 126 \text{ Hz}.$$

For the lower bound (minimum Betas), one obtains:

$$f_L = 9.9 + 2.14 + 58.6 + 2.6 + 40.3 = 113.5 \text{ Hz.}$$
$$\quad C_{11} \quad C_{21} \quad C_{E_1} \quad C_{22} \quad C_{E_2}$$

Noting that the difference between the two extremes is very small, one might conclude that the lower 3dB point is about 120 Hz. However, this is a *conservative* estimate. The actual low-frequency 3dB point of this circuit will lie somewhere between about 100 Hz and 120 Hz. The lower bound of 100 Hz is obtained by averaging the two sums of the contributions of the C_E capacitors. Note that, in all of the above results, these capacitors dominate the low-frequency response. There is no point to further analysis, because the tolerances on these capacitors will surely overcome the accuracy of our estimates. For better accuracy in any "one-shot" design, these capacitors should be measured on a bridge.

Estimates of the High-Frequency 3dB Points

The high-frequency 3dB points of a cascade may be estimated by using equation (3). For the case of our two-stage amplifier, one must consider the effects of both stages.

For the lower bound (maximum Betas), one obtains:

$$\frac{1}{f_H} = \frac{1}{1,014,567} + \frac{1}{269,667}$$
$$\quad \text{1st stage} \quad \text{2nd stage}$$
$$f_H = 119 \text{ KHz.}$$

For the higher bound (minimum Betas), one obtains:

$$\frac{1}{f_H} = \frac{1}{1,435,782} + \frac{1}{269,667}$$
$$f_H = 227 \text{ KHz.}$$

On the average one may expect a high-frequency 3dB point of about 173 KHz. The accuracy of this result could be improved by using equation (4) to calculate C-PI, using the actual operating point; however, in a discrete-component realization the high-frequency 3dB point would also be affected by stray capacitance to ground, and in an integrated circuit, by the substratum capacitance. These additional possible steps are left to the interested user.

This concludes the example of the analysis of multi-stage amplifiers by means of the programs in this book. Other examples will occur to the user. All users are invited to submit their favorite examples to the author for possible inclusion in future editions of this work.

Conversion of Data for a Hybrid Circuit to Data for the Corresponding Hybrid-π Circuit

Modern computer simulation programs use the linear Hybrid-π equivalent circuit for the bipolar junction transistor. Some users may have access to transistor data and circuit-theory equations based on the older Hybrid model; this program may be used to convert Hybrid-model data to the newer Hybrid-π data used in computer programs in this book.

Circuit Description of the Hybrid and Hybrid-π Models

The older Hybrid model for the BJT is illustrated in Figure 26a. [1,2]

---

1. Jacob Millman and Christos Halkias, "Integrated Electronics: Analog and Digital Circuits and Systems", McGraw-Hill Book Company, Inc. 1972.
2. William H. Hayt, Jr. and Gerold W. Neudeck, "Electronic Circuit Analysis and Design", Houghton Mifflin Company, Boston. 1976.

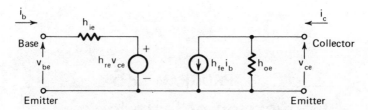

FIGURE 26a. The hybrid model for the bipolar junction transistor.

In this model h_{ie}, h_{re}, and h_{oe} may be frequency-dependent. Because a number of texts currently in use are devoted exclusively to this model, it is necessary to compare it to the newer linear Hybrid-π model.

The linear Hybrid-π model used in our programs is the small-signal equivalent of the more complex non-linear Hybrid-π model found in many of the large-scale computer simulation programs such as SPICE, SLIC and SINC.[3] The model is illustrated in Figure 26b.

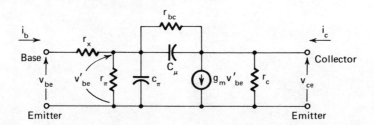

FIGURE 26b. The hybrid π model for the bipolar junction transistor.

The parameters of the linear Hybrid-π model are defined as follows:

C_π = Base-emitter capacitor ⎱ Usually found in
C_μ = Base-collector capacitor ⎰ Transistor Specifications

r_π = Small-signal base-emitter resistance

r_x = Transistor input lead resistance (base to virtual base)

r_c = Small-signal collector-emitter resistance

r_{bc} = Small-signal base-collector resistance

$g_m = |I_c| / V_T$ where $V_T = (T_J)/11,600$

(V$_T$ is the temperature of the junction in electron-volts, and T_J is measured in degrees Kelvin)

h_{fe} = Small-signal Beta, or the small-signal current-gain common-emitter configuration with the collector and emitter shorted

---

3. Ian Getreu, "Modeling the Bipolar Transistor," Tektronix Inc., Beaverton, Oregon, 1976. Part no. 062–2841–00.

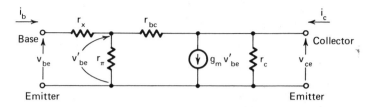

FIGURE 26c. The low frequency hybrid π model for the bipolar junction transistor.

$$h_{fe} = g_m * r_\pi$$

Since the capacitors C_π and C_μ may usually be found in the transistor specifications, it is sufficient to address the problem of comparing the low-frequency limits of the two circuits. In this limit, the Hybrid model of Figure 26a remains unchanged, while the Hybrid-π model of Figure 26b reduces to that of Figure 26c.

In Program 26, the following equations are used to relate the parameters of the Hybrid model to those of the Hybrid-π model:

$$r_\pi = h_{fe}/g_m \qquad r_x = h_{ie} - r_\pi$$
$$r_{bc} = r_\pi/h_{re} \qquad 1/r_c = h_{oe} - (1 + h_{fe}) * h_{re}/r_\pi$$

Computer Procedure:

To convert the Hybrid parameters to the Hybrid-π parameters, the user must INPUT h_{fe}, h_{ie}, h_{re}, h_{oe}, the DC collector current I_c, and the transistor's junction temperature in degrees Centigrade. The program will PRINT the Hybrid-π parameters using the equations above.

Enter Program 26.

Program Listing

```
10 PRINT "HYBRID TO HYBRID PIE
CONVERSION"
20 PRINT
30 INPUT "HFE OR AC BETA= ":HFE
40 PRINT
50 INPUT "HIE= ":HIE
60 PRINT
70 INPUT "HRE= ":HRE
80 PRINT
90 INPUT "HOE= ":HOE
100 PRINT
110 INPUT "DC COLLECTOR CURRENT=
```

```
      ":IC
120 PRINT
130 INPUT "JUNCTION TEMP IN DEG.
 C = ":TJ
140 VT=(TJ+273)/11600
150 GM=IC/VT
160 RPIE=HFE/GM
170 RBC=RPIE/HRE
180 RX=(HIE-RPIE)
190 RC=(1/(HOE-((1+HFE)*HRE/RPIE
)))
200 PRINT
210 PRINT "      THE RESULTS ARE"
220 PRINT
230 PRINT "HFE= ";HFE
240 PRINT
250 PRINT "GM= ";GM
260 PRINT
270 PRINT "R-PIE= ";RPIE
280 PRINT
290 PRINT "RX= ";RX
300 PRINT
310 PRINT "RBC= ";RBC
320 PRINT
330 PRINT "RC= ";RC
340 END
```

Run

```
HYBRID TO HYBRID PIE CONVERSION

HFE OR AC BETA=  61

HIE=  1600

HRE=  .00012

HOE=  .0000058

DC COLLECTOR CURRENT=  .001

JUNCTION TEMP IN DEG. C =  25

     THE RESULTS ARE

HFE=  61

GM=  .0389261745

R-PIE=  1567.068966
```

RX=   32.93103448

RBC=   13058908.05

RC=   950314.7153

Comments: Even though Program A is based on complete theoretical descriptions of the linear Hybrid and the linear Hybrid-π transistor models, the results of some applications may not yield useful results, because some transistor specifications are not self-consistent. Data for the Hybrid parameters obtained from inconsistent specifications may yield a negative value for the Hybrid-π resistor r_x. Very large positive values for r_x are also common. Both types of result are incorrect, and due to inconsistencies in the manufacturer's specifications. This problem is one good reason for using the Hybrid-π description in the circuit theory for the programs in this book, from the outset.

The programs in this book use the Hybrid-π model exclusively. In the majority of programs the user need only specify values for the DC Beta (h_{FE}), the AC Beta (h_{fe}), and the junction temperature, or ambient and overall thermal resistance. High-frequency analysis will also require that the user specify the transistor small-signal internal capacitors such as C_{be} and C_{bc}. These parameters are readily found in transistor specifications, unlike the older Hybrid parameters.

PART II
PROBLEMS

The following problems have been developed to illustrate the use of the programs in both analysis and design. They have been written so as to be of use to the individual user, working alone, or in the classroom or micro-computer laboratory. Each problem includes a printout of the particular program used in its solution and hints for the selection of particular components, as well as extensive suggestions for the analysis and design of similar circuits. The problems should be of considerable value to the student of electronics.

PROBLEM 1 _____

Using the Circuit of Figure 1.1, determine the value of resistor R_B required to yield an operating point of approximately 5 volts for the collector-emitter voltage, V_{CE}.

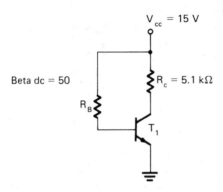

FIGURE 1.1. A fixed bias common emitter stage.

Solution:

Use Program 1. RB MAX has been chosen so that the number of steps, 50 in this case, will yield a number of possibilities. If this circuit is to be realized with 5% resistors, the nearest value would be 360K ohms, yielding an operating point of VCE about 4.8 volts with a collector current, IC, of about 1.99 mA.

Computer Results for Problem 1

```
BJT CE FIXED BIAS-EFFECTS OF
CHANGING RB ON OP POINT

RB MAX=   400000

RC=   5100

VCC=   15

BETA DC=   50

NO OF STEPS=   50

     RB        IC       VCE
   400000    .001793    5.851
   396944    .001807    5.781
   393889    .001821    5.709
   390834    .001835    5.637
```

```
387779    .00185     5.563
384724    .001864    5.488
381669    .001879    5.412
378614    .001895    5.335
375559    .00191     5.256
372504    .001926    5.176
369449    .001942    5.095
366394    .001958    5.012
363339    .001974    4.928
360284    .001991    4.843
357229    .002008    4.756
354173    .002025    4.668
351118    .002043    4.578
348063    .002061    4.486
345008    .002079    4.393
341953    .002098    4.298
338898    .002117    4.202
335843    .002136    4.104
332788    .002156    4.004
329733    .002176    3.902
```

The user should terminate the program upon obtaining the desired range of outputs.

PROBLEM 2 _____

Design the self-biased common-emitter circuit illustrated in Figure 2.1 for a DC operating point of:

$$V_{CE} = 7 \text{ volts}, \qquad I_c = 2.0 \text{ mA} \pm 5\%.$$

Use the 2N4123 transistor and a power supply of 20 volts.

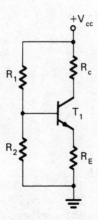

FIGURE 2.1. A self biased common emitter stage.

Solution:

Use Program 4. The bounds on the collector current are ± 5%; hence I_c may range from a high of 2.1 mA to a low of 1.9 mA. Referring to the specification sheets for the 2N4123, note that the DC Beta, h_{FE}, ranges from 50 to 150 for a collector current of 2 mA. The reverse saturation current (common-base), ICBO, has a maximum value of 50 nA. We elect to INPUT the average value for ICEO, namely ICEO = 100*50E-9, or 5 micro-amps. We are now ready to RUN Program 4.

Computer Results for Problem 2

```
PGM WILL DESIGN A CE-BJT DC SELF
-BIASED CIRCUIT FOR A   GIVEN DE
SIGN TARGET.

USER MUST SPECIFY TARGET    V-CE
 AND RANGE OF I-C.

VCC=  20
TARGET V-CE=  7
UPPER BOUND IC=   .0021
LOWER BOUND IC=   .0019

BETA DC MAX=  150
BETA DC MIN=  50

I-CEO=   .000005

R1=  60621.59606

R2=  43903.69957

RE=  3500

RC=  3000

VCE MAX=  7.775825862

VCE MIN=  6.405560929

IC MAX=  .0020989369

IC MIN=  .00190071

DO YOU WANT TO CHECK YOUR DESIGN
 FOR ITS OPERATING RANGE WHEN YO
U SPECIFY RESISTORS?

IF YES ENTER 1, IF NO ENTER 0.

A=  1
```

Rather than terminate this RUN by INPUTing a 0, continue with Problem 3 and finalize the design by choosing 5% resistors and calculating the actual operating point.

PROBLEM 3 _____

Using the results of the design of Problem 2, finalize the design by choosing 5% resistors and calculating the actual operating point of the circuit. Refer to Figure 2.1.

Solution:

This is a continuation of the application of Program 4 to solve Problem 2. We begin by using the nearest 5% resistor for all components. Note that the choice of RE = 3600 ohms results in an operating point that is somewhat low. This may be corrected by decreasing RE to 3300 ohms, as in the second computer RUN below. The second RUN is good enough and should yield the desired operating point for any transistor that meets the specifications for the population.

Computer Results for Problem 3

Run # 1

```
R1=   62000

R2=   43000

RE=   3600

RC=   3000

ICMAX= .0019881113

ICMIN= .0018047175

VCE MAX= 8.216256568

VCE MIN= 6.925864266

DOES YOUR 5,10,OR20% RESISTOR DE
SIGN MEET SPECS ?

IF NOT ENTER 1 TO RE-SPECIFY RES
ISTORS

IF SPECS. ARE MET THEN ENTER 0 T
O END PROGRAM.
```

Run # 2

```
B=   1
R1=   62000

R2=   43000

RE=   3300

RC=   3000

ICMAX= .002160083

ICMIN= .001947268

VCE MAX= 7.858211432

VCE MIN= 6.438684379

DOES YOUR 5,10,OR20% RESISTOR DE
SIGN MEET SPECS ?

IF NOT ENTER 1 TO RE-SPECIFY RES
ISTORS

IF SPECS. ARE MET THEN ENTER 0 T
O END PROGRAM.

B=   0
```

PROBLEM 4 ⎯⎯⎯

Design a common-emitter amplifier using a 2N4123 transistor, to the following specifications:

Overall Mid-Band Voltage Gain, $A_{V.S} \geq 150$.
Mid-Band Input Resistance, $Z_{in} \geq 500$ ohms.

The source impedance is 500 ohms and a 20-volt power supply is available. The circuit is illustrated in Figure 4.1.

Solution:

Use Program 5. Assume a junction temperature equal to the typical ambient temperature of 25°C. For a voltage amplifier, such an approximation is reasonable. The sample computer RUN that follows is self-explanatory; however, some hints are in order.

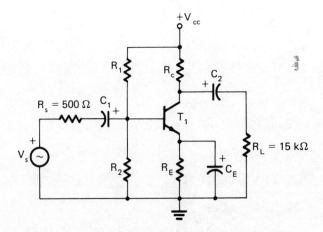

FIGURE 4.1. A single stage common emitter linear amplifier.

Note that the DC collector current is targeted for 1.6 mA. The transistor's Beta specification for 2 mA is close enough for this example.

We have also elected to use a design parameter GAMMA of 40. If more stability against transistor-population variations in Beta is desired, the user should increase GAMMA.

Computer Results for Problem 4

```
COMMON EMMITER, MID-  BAND VOLTA
GE AMPLIFIER DESIGN
ALL R IN OHMS
ALL I IN AMPS
ALL V IN VOLTS

ENTER MINIMUM ALLOWABLE VOLTAGE
GAIN AT MIDBAND
AV MIN=   150

ENTER MINIMUM ALLOWABLE INPUT IM
PEDANCE
ZIN MIN=   500

ENTER SOURCE IMPEDANCE
ZS=   500

ENTER POWER SUPPLY VOLTAGE
VCC=   20

ENTER JUNCTION TEMP. IN DEG. C
TJ=   25

ENTER LOAD RESISTOR
```

RL= 15000

ENTER COLLECTOR RESISTOR
RC= 7500

THE LOWER BOUND ON GM IS GM= .06

THE LOWER BOUND ON THE COLLECTOR
 CURRENT IS, IC= .0015413793

CHOOSE A DESIGN VALUE FOR THE CO
LLECTOR CURRENT
I

ENTER MINIMUM TRANSISTOR SMALL S
IGNAL BETA AT THE OPERATING POIN
T
MIN BETA AC= 50

THE SMALL SIGNAL TRANSISTOR INPU
T IMPEDANCE R-PIE= 802.8017241

ENTER MINIMUM D.C.BETA AT THE OP
ERATION POINT
MIN BETA DC= 50

DESIGNER MUST ENTER PARAMETER GA
MMA
GAMMA= 40

ZIN= 626.0124567

R1= 11992.1875

R2= 3725.961538

RE= 2500

RC= 7500

RL= 15000

AVS= 166.7865536

THIS COMPLETES YOUR COMPUTER DES
IGN. TO REPEAT FOR DIFFERENT GAM
MA, ENTER 1, TO END ENTER 0.
X= 0

Design is not unique, so the user's results may differ from the above. Note that an available 5% resistor is used for R_c.

PROBLEM 5 _____

Design a low-level common-emitter amplifier to operate at a remote site as part of a sensor interface for a micro-computer. The sensor impedance is 500 ohms. A 9-volt battery is suggested as the power supply. The overall mid-band voltage gain into a 15K-ohm load must be $\geq$ 100. (Note that the load could be the input impedance of an emitter-follower used to drive the bus to the interface.) A 2N3711 transistor is suggested. Design the circuit for high Beta stability. Use the schematic of Figure 4.1.

Solution:

An AVS of 100 is difficult at 9 volts. Proceed by keeping Z_{in} high, say 3K ohms or more. The DC collector current is targeted at 0.8 mA; use the Beta specifications for 1.0 mA. Two values of GAMMA are used, so as to illustrate the differences in the designs.

Computer Results for Problem 5

```
COMMON EMMITER, MID-  BAND VOLTA
GE AMPLIFIER DESIGN
ALL R IN OHMS
ALL I IN AMPS
ALL V IN VOLTS

ENTER MINIMUM ALLOWABLE VOLTAGE
GAIN AT MIDBAND
AV MIN=   100

ENTER MINIMUM ALLOWABLE INPUT IM
PEDANCE
ZIN MIN=   3000

ENTER SOURCE IMPEDANCE
ZS=   500

ENTER POWER SUPPLY VOLTAGE
VCC=   9

ENTER JUNCTION TEMP. IN DEG. C
TJ=   25

ENTER LOAD RESISTOR
RL=   15000

ENTER COLLECTOR RESISTOR
```

RC=   5100

THE LOWER BOUND ON GM IS GM=
.0306535948

THE LOWER BOUND ON THE COLLECTOR
 CURRENT IS, IC= .0007874803

CHOOSE A DESIGN VALUE FOR THE CO
LLECTOR CURRENT
IC=   .0008

ENTER MINIMUM TRANSISTOR SMALL S
IGNAL BETA AT THE OPERATING POIN
T
MIN BETA AC=   180

THE SMALL SIGNAL TRANSISTOR INPU
T IMPEDANCE R-PIE=   5780.172414

ENTER MINIMUM D.C.BETA AT THE OP
ERATION POINT
MIN BETA DC=   180

DESIGNER MUST ENTER PARAMETER GA
MMA
GAMMA=   70

ZIN= 3082.309172

R1= 18932.14286

R2= 10141.30435

RE= 3075

RC= 5100

RL= 15000

AVS= 100.3829428

THIS COMPLETES YOUR COMPUTER DES
IGN. TO REPEAT FOR DIFFERENT GAM
MA, ENTER 1, TO END ENTER 0.
X=   1

DESIGNER MUST ENTER PARAMETER GA
MMA
GAMMA=   50

```
ZIN= 3561.774137

R1= 26505

R2= 14280.61224

RE= 3075

RC= 5100

RL= 15000

AVS= 102.3051263

THIS COMPLETES YOUR COMPUTER DES
IGN. TO REPEAT FOR DIFFERENT GAM
MA, ENTER 1, TO END ENTER 0.
X=   0
```

PROBLEM 6 _____

Design a low-level common-emitter amplifier to match a 5K-ohm source to a 5K-ohm load, with a mid-band gain of 50. Use a 2N3711 transistor. A 15-volt power supply is available. The circuit is illustrated by Figure 4.1, except that the values of RS and RL must be changed to 5K ohms.

Solution:

The comments and hints are similar to those of Problems 4 and 5.

Computer Results for Problem 6

```
COMMON EMMITER, MID-  BAND VOLTA
GE AMPLIFIER DESIGN
ALL R IN OHMS
ALL I IN AMPS
ALL V IN VOLTS

ENTER MINIMUM ALLOWABLE VOLTAGE
GAIN AT MIDBAND
AV MIN=   50

ENTER MINIMUM ALLOWABLE INPUT IM
PEDANCE
ZIN MIN=   5000

ENTER SOURCE IMPEDANCE
ZS=   5000
```

```
ENTER POWER SUPPLY VOLTAGE
VCC=  15

ENTER JUNCTION TEMP. IN DEG. C
TJ=  25

ENTER LOAD RESISTOR
RL=  5000

ENTER COLLECTOR RESISTOR
RC=  12000
THE LOWER BOUND ON GM IS GM=
 .0283333333

THE LOWER BOUND ON THE COLLECTOR
 CURRENT IS, IC= .0007278736

CHOOSE A DESIGN VALUE FOR THE CO
LLECTOR CURRENT
IC=  .00073

ENTER MINIMUM TRANSISTOR SMALL S
IGNAL BETA AT THE OPERATING POIN
T
MIN BETA AC=  180

THE SMALL SIGNAL TRANSISTOR INPU
T IMPEDANCE R-PIE=  6334.435522


ENTER MINIMUM D.C.BETA AT THE OP
ERATION POINT
MIN BETA DC=  180

DESIGNER MUST ENTER PARAMETER GA
MMA
GAMMA=  20

ZIN= 5389.989465


R1= 138452.0548

R2= 48925.73901

RE= 4273.972603

RC= 12000

RL= 5000
```

```
AVS= 51.87675583

THIS COMPLETES YOUR COMPUTER DES
IGN. TO REPEAT FOR DIFFERENT GAM
MA, ENTER 1, TO END ENTER 0.
X=   0
```

PROBLEM 7 _____

Analyze the results of your design for Problem 5 for: the DC operating point; the input impedance; and the overall voltage gain at mid-band. Find the range of the above for the range of DC and AC Beta's given in the transistor's specifications.

Solution:

Use Program 6. The procedure is straightforward. Because the expected collector current is about 0.8 mA, use the data available for a current of 1 mA. If the expected collector current had not been near a value listed in the specification sheet for the transistor, it would have been necessary to consult the specifications for Chip N14 and correct for the different IC. The "average" value for ICEO is used. This is obtained from the maximum value for ICBO and an averge value for DC Beta, namely 420.

Computer Results for Problem 7

```
ANALYSIS OF A COMMON EMITTER BJT
 AMPLIFIER FOR RANGE OF D.C. AND
 A.C. OP.

R1=   18000

R2=   10000

RE=   3000

RC=   5100

RL=   15000

RS=   500

VCC=  9

BETA DC MAX=   660
BETA DC MIN=   180

BETA AC MAX=   800
```

```
BETA AC MIN=   180

TEMP DEG. C=   25

I-CEO=   .000042

ICMAX=   .00085091
ICMIN=   .0008408144

VCE MAX=   2.203339894
VCE MIN=   2.111491305

ZIN MAX=   5077.201587
ZIN MIN=   2963.95406

AVS MAX=   114.7621752
AVS MIN=   106.5875706
```

PROBLEM 8 _____

Analyze the results of your design for Problem 4. Finalize the design by selecting 5% resistors, and calculate the ranges for IC, VCE, ZIN, and AVS, given the ranges for the transistor's DC and AC Beta's.

Solution:

Use Program 6. The procedure is straightforward. Because the design is targeted for a collector current of about 1.6 mA, use the Beta data for 2 mA. A value of the "average" reverse saturation current (common-emitter) is obtained as in Problem 2.

Computer Results for Problem 8

```
ANALYSIS OF A COMMON EMITTER BJT
 AMPLIFIER FOR RANGE OF D.C. AND
 A.C. OP.

R1=   12000

R2=   3600

RE=   2400

RC=   7500

RL=   15000

RS=   500
```

```
VCC=   20

BETA DC MAX=   150
BETA DC MIN=   50

BETA AC MAX=   200
BETA AC MIN=   50

TEMP DEG. C=   25

I-CED=  .000005

ICMAX=  .0016289257
ICMIN=  .0015842158

VCE MAX=   4.390814608
VCE MIN=   3.899525953

ZIN MAX=   1474.59928
ZIN MIN=   627.1714445

AVS MAX=   236.759843
AVS MIN=   171.5625042
```

PROBLEM 9 _____

Analyze the results of your design for Problem 6. Finalize the design by selecting 5% resistors and calculate the ranges for IC, VCE, ZIN and AVS, given the ranges for the transistor's DC and AC Beta's.

Solution:

Use Program 6. Refer to Problem 7 for suggestions.

Computer Results for Problem 9

```
ANALYSIS OF A COMMON EMITTER BJT
 AMPLIFIER FOR RANGE OF D.C. AND
 A.C. OP.

R1=   130000

R2=   47000

RE=   4300

RC=   12000

RL=   5000

RS=   5000
```

```
VCC=  15

BETA DC MAX=  660
BETA DC MIN=  180

BETA AC MAX=  800
BETA AC MIN=  180

TEMP DEG. C=  25

I-CEO=  .000042

ICMAX=  .0007652353
ICMIN=  .0007401144

VCE MAX=  2.953718835
VCE MIN=  2.531642991

ZIN MAX=  15104.9402
ZIN MIN=  5290.34856

AVS MAX=  78.98693538
AVS MIN=  52.27536614
```

PROBLEM 10 _____

Design a common-base low-level amplifier with a mid-band input impedance of about 52 ohms and an output impedance of about 6K ohms. A 9-volt power supply is available. Use the 2N3711 transistor. The circuit is illustrated in Figure 10.1.

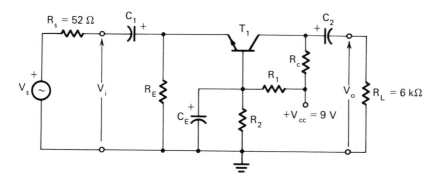

FIGURE 10.1. A single stage common base linear amplifier.

Solution:

Use Programs 4 and 7. The DC operating points of a common-base and a common-emitter circuit are identical, so use Program 4 to design for a particular DC operating point.

The input impedance of a common-base amplifier at mid-band is given by equation (1) of Program 7. We note that ZIN is controlled by the transconductance, g_m, since g_m will be much larger than either $1/RE$ or $1/r_\pi$. In effect, ZIN is determined by these three "resistances" in parallel. We proceed by targeting the $1/g_m$ of our circuit to be about 55 ohms. As $g_m = I_c/V_T$, or about $38.9 * I_c$ at 25°C, design for a collector current of about 0.46 mA.

Computer Results for Problem 10 (from Program 4)

```
PGM WILL DESIGN A CE-BJT DC SELF
-BIASED CIRCUIT FOR A   GIVEN DE
SIGN TARGET.

USER MUST SPECIFY TARGET    V-CE
 AND RANGE OF I-C.

VCC=  9
TARGET V-CE=  3
UPPER BOUND IC=  .00047
LOWER BOUND IC=  .00046

BETA DC MAX=  660
BETA DC MIN=  180

I-CEO=  0

R1=  69175.10171

R2=  48448.3874

RE=  6451.612903

RC=  6451.612903

VCE MAX=  3.080912493

VCE MIN=  2.940266311

IC MAX=  .0004699849

IC MIN=  .00046

DO YOU WANT TO CHECK YOUR DESIGN
 FOR ITS OPERATING RANGE WHEN YO
U SPECIFY RESISTORS?

IF YES ENTER 1, IF NO ENTER 0.

A=  0
```

This is the desired DC design, so proceed by choosing 5% resistors and using Program 7. The computer results follow. Note that we have met our original specifications very well, with ZIN varying from about 52 to 53 ohms for any transistor, despite the large Beta range. The voltage gain and power gain are also good for a single stage operating with a power supply of 9 volts. The output impedance for a common-base amplifier is given simply by RC. This circuit could be used as a remote amplifier, matched to a coaxial cable of 52 ohms.

Computer Results for Problem 10 (from Program 7)

```
ANALYSIS OF A COMMON BASE BJT AM
PLIFIER FOR RANGE OF D.C. AND A.
C. OP.

R1=   68000

R2=   47000

RE=   6200

RC=   6200

RL=   6000

RS=   52

VCC=   9

BETA DC MAX=   660
BETA DC MIN=   180

BETA AC MAX=   800
BETA AC MIN=   180

TEMP DEG. C=   25

I-CEO=   .000042

ICMAX=   .0004847513
ICMIN=   .0004752337

VCE MAX=   3.123381179
VCE MIN=   2.993630234

ZIN MAX=   52.48133753
ZIN MIN=   53.29612104

AVI MAX=   57.53655455
AVI MIN=   56.40687493
```

```
AVS MAX=  28.90081053
AVS MIN=  28.5506019

MAX CURRENT GAIN=  .50326589
MIN CURRENT GAIN=  .5010446057

MAX POWER GAIN=  28.95618533
MIN POWER GAIN=  28.26236041

OUTPUT Z=  6200
```

PROBLEM 11 _____

Design an emitter-follower low-level amplifier with a mid-band output impedance of about 52 ohms. A 9-volt power supply is available. Use the 2N4123 transistor. The circuit is illustrated in Figure 11.1.

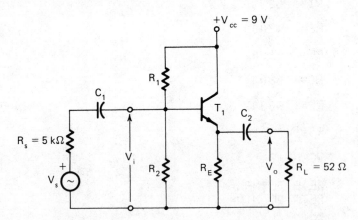

FIGURE 11.1. A single stage emitter follower linear amplifier.

Solution:

Use Programs 20 and 8. Prior to using Program 20 for the design, line number 95 should be changed to VO=.65 since this design is a voltage amplifier. The total thermal resistance, junction to ambient, may be obtained from the transistor specifications by way of Notes 2, and the TI guaranteed curve. RJA = 125/625 = 0.2°C per milliwatt. Since programs require that this be entered as degrees C per watt, we must use 200 here. Once the design has been completed by way of Program 20, we proceed with the complete analysis using Program 8.

Computer Results for Problem 11

```
POWER AMP DESIGN

TARGET VCE=  5.7
```

```
TARGET IC=  .0007
TARGET ZO=  52


VCC=  9
RS=  5000
AMBIENT TEMP=  25
TOTAL THERMAL RESISTANCE-JUNCTIO
N TO AMBIENT=  200
BETA DC=  100
BETA AC=  125
INITIAL GAMMA=  100


DESIGN GAMMA=  97
VCE=  5.732449515
IC=  .0007
VBE=  .65
TJ=  25.79465138
DC POWER IN=  .0039732569
GM=  .0269086293
RPIE=  4645.34995
R1=  7437.407953
R2=  5877.97619
RE=  4714.285714


ZO=  52.01639155
ZIN=  2535.199105
AVI=  .5825488367
AP=  16.54003904


ANALYSIS OF AN EMITTER FOLLOWER
BJT AMPLIFIER FOR RANGE OF D.C.
AND A.C. OP


R1=  7500

R2=  5600

RE=  4700

RL=  52

RS=  5000

VCC=  9

BETA DC MAX=  150
BETA DC MIN=  50

BETA AC MAX=  200
BETA AC MIN=  50
```

```
AMBIENT TEMP IN DEG C =   25

THERMAL RESISTANCE, JUNCTION TO
AMBIENT, IN DEG. C/WATT=   200

I-CEO=   0

ICMAX=   .0006727383
ICMIN=   .0006581408

VCE MAX=   5.96739057
VCE MIN=   5.859069564

T-J MAX DEG C = 25.78832409
T-J MIN DEG C = 25.7854766

ZIN MAX=   2721.271859
ZIN MIN=   1885.88678

AVI MAX=   .5744667218
AVI MIN=   .5727270195

AVS MAX=   .2024640697
AVS MIN=   .1568568216

MAX CURRENT GAIN=   30.06307931
MIN CURRENT GAIN=   20.77112143

MAX POWER GAIN=   17.27023862
MIN POWER GAIN=   11.89618247

MAX OUTPUT Z=   75.44215801
MIN OUTPUT Z=   47.33451563
```

PROBLEM 12 _____

Using your design for the common-emitter amplifier of Problem 6, select capacitors C_1, C_2, and C_E so that the low-frequency 3dB point will be about 50 Hz. In addition, calculate the high-frequency 3dB point at the actual operating point of the circuit.

Solution:

The schematic for the CE amplifier is illustrated in Figure 12.1, using the results of Problem 6 and selecting 5% resistors.

Note that, although R_s is large, it is typical of the output impedance for a driving CE amplifier, in this case R_c of the previous stage.

Program 9 may be used to solve the problem as follows. In order to minimize the value of capacitor C_E, let this capacitor control the low frequency 3dB point, f_L. Begin with a trial of, say, $C_1 = C_2 = 1\mu F$, $C_E = 100\mu F$.

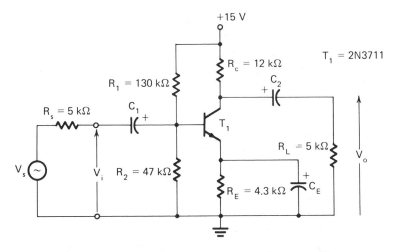

FIGURE 12.1. A single stage common emitter linear amplifier.

Next, develop some typcial values for C-PI and C-MU at the operating point of the circuit. Because the transistor is operating in its forward region, C-MU is a depletion capacitance, and is given in Figure 14 of the specification sheets for Chip Type 21. We require only the voltage V_{CB}. From our results for Problem 9, where this circuit was analyzed previously, we note that $V_{CE} \cong 2.7$ volts. Since $V_{BE} = 0.65$ volts, it follows that $V_{BC} \cong 2$ volts. If the user has not done Problem 9, then Program 9 may be used to calculate the DC operating point of the circuit with a 5% realization for the resistors. From Figure 14 in the specification sheet, C-MU or C_{cb} has a value of 5 pF.

Now calculate the value for C-PI from the gain-bandwidth product ω_T. Since C-PI is a diffusion capacitance for the forward-biased base-emitter junction, its value may be calculated from the equation:

$$\text{C-PI} = \frac{g_m}{\omega_T} - \text{C-MU}.$$

Using the average value of the DC operating point yields a collector current of 0.75 mA. Figure 13 of the specification sheets shows a value for f_T of about 85 mHz. Hence at a junction temperature of about 25°C:

$$g_m = \frac{I_c}{V_T} \quad , \quad V_T = \frac{T°\,K}{11,600}$$

to yield $g_m = 29.175$ and C-PI = 49.6 pF. These values are INPUTed when required.

For Beta AC and Beta DC, INPUT the values in the specification sheets for the 2N3711 transistor at a collector current of 1 mA, because this is very close to the actual operating value.

Next, develop a value for the junction-to-ambient thermal resistance for the transistor in the absence of a heat sink. This may be calculated from Figure 1 of the specification sheet for the 2N3711 as follows:

$$TR = \frac{150 - 25}{625} \times 10^3 = 200 \text{ C}°/\text{watt}.$$

The results of our computer run indicate that FC1MAX = 15.4 Hz, FC2MAX = 9.36 Hz and FCEMAX = 41.1 Hz. Although these results appear to meet the specification for the low-frequency 3dB point, it is best to require that the sum of the three 3dB points be less then 50 Hz. Therefore, repeat the low-frequency analysis, increasing C_1 to $4\mu F$ and C_2 to $3\mu F$. * The results of the computer analysis indicate that the design should meet the low-frequency specification of 50 Hz.

Note that the high-frequency 3dB point will be at least 72.55 KHz, ranging from this low point to a high of about 109 KHz. Any actual stray capacitance in the final physical realization will act to decrease this result. This completes our discussion of the solution.

Computer Results for Problem 12

```
COMPLETE ANALYSIS OF A COMMON EM
ITTER AMPLIFIER

R1=   130000

R2=   47000

RE=   4300

RC=   12000

RL=   5000

RS=   5000

C1=   .000001

C2=   .000001

CE=   .0001

C PIE=   4.96E-11

C MU=   5.E-12
```

---

*Available from SPRAGUE, TE Miniature Electrolytics

```
VCC=  15
BETA DC MAX=  660
BETA DC MIN=  180

BETA AC MAX=  800
BETA AC MIN=  180

AMBIENT TEMP. IN DEG C = 25

THERMAL RESISTANCE, JUNCTION TO
AMBIENT, IN DEG. C/WATT =  200

I-CEO=  0

ICMAX=  .0007646685
ICMIN=  .0007381085

VCE MAX=  2.986367258
VCE MIN=  2.540877289

T-J MAX DEG C = 25.44085259
T-J MIN DEG C = 25.38858578

ZIN MAX=  15123.79559
ZIN MIN=  5308.36834

AVS MAX=  78.83628798
AVS MIN=  52.15181384

AVI MAX=  104.8999462
AVI MIN=  101.2740775


MAX CURRENT GAIN=  317.2970688
MIN CURRENT GAIN=  107.5200213

MAX POWER GAIN=  33284.44546
MIN POWER GAIN=  10888.99097

OUTPUT Z=  12000

F C1 MAX=  15.43939309
F C1 MIN=  7.908793855

F C2=  9.362055934

F CE MAX=  41.12063595
F CE MIN=  27.88718031

MAX H.F. 3 D.B. POINT=
  108859.2007
```

First design attempt for f_L.

```
MIN H.F. 3 D.B. POINT=
  72550.12151


C1=  .000004

C2=  .000003

CE=  .0001

F C1 MAX=  3.859848271
F C1 MIN=  1.977198464

F C2=  3.120685311

F CE MAX=  41.12063595
F CE MIN=  27.88718031
```

Second design attempt for f_L.

PROBLEM 13

Run a quick analysis for the low- and high-frequency 3dB points of the circuit of Figure 12.1 when $C_1 = 8$ F, $C_2 = 6$ F, and $C_E = 200$ F. Use the average value of the extremes of the DC operating point, as calculated in Problem 9. Estimate the effect of a 10 pF stray capacitance from collector to ground on the high-frequency 3dB point.

Solution:

Use Program 10. The required calculations for C-PI and C-MU are similar to those of Problem 12 and will not be repeated. They yield:

$$C\text{-}PI = 49.6 \text{ pF}; \qquad C\text{-}MU = 5 \text{ pF};$$
$$IC_{av} = 0.75 \text{ mA}; \text{ and} \qquad BETA \ AC_{av} = 490.$$

Computer Results for Problem 13

```
A SIMPLIFIED LOW AND HIGH FREQ.
ANALYSIS FOR THE COMMON EMITTER
AMPLIFIER.

R1=  130000

R2=  47000

RE=  4300

RC=  12000

RL= 5000
```

```
RS=  5000

C1=  .000008

C2=  .000006

CE=  .0002

C PIE=  4.96E-11

C MU=  5.E-12

I-C D.C.=  .00075

BETA A.C.=  490

THE LOW FREQUENCY 3 DB POINT IS
DETERMINED BY THE LARGER OF THE
FOLLOWING THREE FREQUENCIES IN H
Z:
F C1=  1.221023975

F C2=  1.560342656

F CE=  18.65773444

THE HIGH FREQUENCY 3 DB POINT IS
=  79883.87182

TO DO STRAY CAP. ENTER 1, TO END
 ENTER 0

CS=  1.E-11
H.F. 3 D.B. DUE TO CS=
 4509390.275
```

It is clear that the high-frequency 3dB point is controlled by the Miller Effect and not by the 10-pF stray capacitance.

PROBLEM 14 _____

Design a common-source voltage amplifier using the 2N6451 N-JFET to operate between a 52-ohm source and a 20K-ohm load. A 15-volt power supply is available. Calculate all mid-band gains and the DC operating points at the transistor's population extremes for IDSS and VP.

Solution:

This is not an easy design, because it requires that the circuit operate properly for any transistor sample satisfying the manufacturer's specifications.

Begin by designing for a DC operating point using Program 11, using the average values for IDSS and VP, namely 12.5 mA and − 2 volts. In view of the large differences between the extremes in IDSS, 5 and 20 mA, target the average value for ID at 6 mA, and VDS at about half of VDD.

Computer Results from Program 11

```
DESIGN OF A COMMON SOURCE AMPLI-
FIER FOR A GIVEN DC OPERATING
POINT

VDD=  15
TARGET VDS=  8
TARGET ID=  .006
IDSS=  .0125
VP= -2
VGS= -.6143593539

CHOOSE R= 200
CHOOSE R2=  51000

R1=  1201760.849
R2=  51000
RD=  966.6666667
R=  200
A=  0
```

Having the DC design, select 5% resistors and proceed with the required circuit analysis for all mid-band and DC operating points. These may be obtained by using Program 12.

Conclude by noting that the original design does indeed meet the specifications imposed on the circuit by the large spread of IDSS and VP so typical of field-effect transistors. The circuit should work properly, and operate in the saturation region, for any transistor in the sample population. The circuit's overall voltage gain AVS will range from a possible low of 7.25 to a possible high of 15.8. Low voltage gains are typical of FET's.

Computer Results for Problem 14

```
ANALYSIS OF A CS NJFET FOR D.C.
OP. POINT & ALL MID-BAND GAINS

R1=  1200000
R2=  51000
RD=  1000
R=  200
RS= 52
```

```
RL= 20000
VDD= 15
IDSS=  .02
VP= -3.5

VGS1= -1.166508306
VGS2= -8.895991694

CHOOSE CORRECT VGS. CORRECT ABS(
VGS) < ABS(VP).
VGS= -1.16651

ID=  .0088900826
VDS=  4.331900904

GM= .0076195592
AVI=-7.256723032
AVS=-7.249017753
CURRENT GAIN=-17.75025778
POWER GAIN= 128.8087044

ZIN= 48920.86331
ZOUT=  1000
TO REPEAT FOR ANOTHER IDSS & VP
ENTER 1. TO END ENTER 0
X= 1
IDSS=  .005
VP= -.5

VGS1= -.0832411073
VGS2= -1.166758893

CHOOSE CORRECT VGS. CORRECT ABS(
VGS) < ABS(VP).
VGS= -.08324

ID=  .003473778
VDS=  10.83146646

GM= .0166704
AVI=-15.87657143
AVS=-15.85971349
CURRENT GAIN=-38.83477903
POWER GAIN= 616.5631432

ZIN= 48920.86331
ZOUT=  1000

TO REPEAT FOR ANOTHER IDSS & VP
ENTER 1. TO END ENTER 0
X= 0
```

PROBLEM 15 _____

Using the results of Problem 14 for the common-source voltage amplifier, complete the design by choosing a set of coupling capacitors C_1 and C_2 and a by-pass capacitor C_s so that the average 3dB low-frequency point for the population will be about 50 Hz. Calculate the spread for both the low- and high-frequency 3dB points at the actual extreme operating points of the circuit.

Solution:

Use Program 13. This program may easily be repeated by using the branching command, so begin by taking an educated guess for the capacitors: $C_1 = C_2 = 1\mu F$, and $C_s = 50\mu F$.

The circuit operating point extremes have already been calculated in Problem 14 for the extremes of IDSS and VP, as given by the specifications. Beginning with the high current extreme of IDSS = 20 mA and VP = −3.5 volts, note that VGS ≅ −1 volt. VGS is required to obtain values for the transistor capacitances, C-GS and C-GD. From Figures 6 and 7 of the specifications for Chip JN55:

$$C_{rss} = 3.4\,pF = C_{gd}, \text{ and}$$

$$C_{gs} = C_{iss} - C_{gd} = 14 - 3.4 = 11.6\,pF.$$

Next, note that the 3dB frequency is reasonable; however, FC2 is more than 10% of 40 Hz. Therefore, use branching code "1" and repeat the analysis. Increasing capacitor C_2 to $2\,\mu F$ yields a more reasonable set of results for the 3dB points of the three capacitors. Now use branching code "2".

The low current extreme is given by IDSS = 5 mA and VP = −.5 volts resulting in a VGS of −0.1 volts (see the results of Problem 14). INPUTing the correct value for the transconductance produces the completed results. Note that the low-current population will have the minimum bandwidth, because they yield the largest low-frequency 3dB points and the smallest high-frequency 3dB points.

We may conclude that our design meets the following specifications:

$$7.25 \leq \text{Overall Voltage Gain} \leq 15.86$$

$$40 \leq \text{Low-Frequency 3dB Point} \leq 69$$

$$21.7\,\text{MHz} \leq \text{High-Frequency 3dB Point} \leq 30\,\text{MHz}.$$

Computer Results for Problem 15

```
A SIMPLIFIED LOW AND HIGH FREQUE
NCY ANALYSIS FOR THE COMMON SOUR
CE AMPLIFIER
```

```
R1=   1200000

R2=   51000

RD=   1000

R=   200

RS=   52

RL=   20000

C1=   .000001

C2=   .000001

CS=   .00005

C-GS=   1.16E-11

C-GD=   3.4E-12

GM AT OP POINT=   .00762

THE LOW FREQUENCY 3 DB POINT IS
THE LARGER OF THE FOLLOWING THRE
E FREQUENCIES IN HZ:
F-C1=   3.249860027
F-C2=   7.578807184
F-CS=   40.1707096

THE HIGH FREQUENCY 3 DB POINT IS
=   30035087.14

A=   1

C1=   .000001

C2=   .000002

CS=   .00005

THE LOW FREQUENCY 3 DB POINT IS
THE LARGER OF THE FOLLOWING THRE
E FREQUENCIES IN HZ:
F-C1=   3.249860027
F-C2=   3.789403592
F-CS=   40.1707096

A=   2
```

C-GS= 1.44E-11

C-GD= 3.6E-12

GM AT OP POINT= .01667

THE LOW FREQUENCY 3 DB POINT IS
THE LARGER OF THE FOLLOWING THRE
E FREQUENCIES IN HZ:
F-C1= 3.249860027
F-C2= 3.789403592
F-CS= 68.97775571

THE HIGH FREQUENCY 3 DB POINT IS
= 21705578.07

A= 0
THIS COMPLETES THE ANALYSIS

PROBLEM 16 _____

Design a common-gate voltage amplifier with a mid-band input impedance
of about 50 ohms. Use a 2N6451 N-JFET transistor, a 15-volt power supply,
and a load resistor of 20K ohms. This circuit will eventually form the basis of
a very wide-band amplifier. Analyze your design for all mid-band gains and
for the input impedances at the population extremes, so as to establish the
range of the mid-band gains and the range of the input impedances.

Solution:

Because the resistor R of our design for the common-source amplifier of
Problem 14 was 200 ohms, that DC design should work rather well here also.
The proposed schematic is illustrated by Figure 16.1.

Check the design by using Program 14. INPUTing the resistors with a
5% realization and using the high-current extremes yields an overall voltage

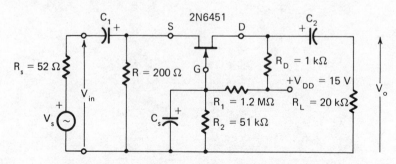

FIGURE 16.1. A single stage common base linear amplifier.

gain of 4.38 and an input impedance of 79 ohms. The low-current extreme yields a gain of 7.4 and an input impedance of 46 ohms. This low voltage gain is typical of very wide-band amplifiers. The range could be decreased by increasing R_2 and R and decreasing R_D, but then the circuit would produce even less voltage gain, unless a larger power supply were used.

Computer Results for Problem 16

```
ANALYSIS OF A CB NJFET FOR D.C.
OP. POINT & ALL MID-BAND GAINS

R1=  1200000
R2=  51000
RD=  1000
R=  200
RS= 52
RL= 20000
VDD= 15
IDSS=  .02
VP= -3.5

VGS1= -1.166508306
VGS2= -8.895991694


CHOOSE CORRECT VGS. CORRECT ABS(
VGS) < ABS(VP).
VGS= -1.16651

ID=  .0088900826
VDS=  4.331900904

GM= .0076195592
AVI= 7.256723032
AVS= 4.381504774
CURRENT GAIN= .0287518879
POWER GAIN= .208644875

ZIN= 79.24207062
ZOUT= 1000


TO REPEAT FOR ANOTHER IDSS & VP
ENTER 1. TO END ENTER 0
X= 1
IDSS=  .005
VP= -.5

VGS1= -.0832411073
VGS2= -1.166758893
```

```
CHOOSE CORRECT VGS. CORRECT ABS(
VGS) < ABS(VP).
VGS= -.08324

ID=  .003473778
VDS=  10.83146646

GM= .0166704
AVI= 15.87657143
AVS= 7.464791033
CURRENT GAIN= .0366319298
POWER GAIN= .58158945

ZIN= 46.14589486
ZOUT= 1000

TO REPEAT FOR ANOTHER IDSS & VP
ENTER 1. TO END ENTER 0
X= 0
```

PROBLEM 17 ⎯⎯⎯

Using the results of Problem 16 for the common-gate voltage amplifier, complete the design by choosing a set of coupling capacitors C_1 and C_2 and a by-pass capacitor C_s which will make the average low-frequency 3dB point about 50 Hz. Calculate the spreads for both the low- and high-frequency 3dB points at the actual extreme operating points of the circuit.

Solution:

Use Program 15. The program may easily be repeated by using the branching command; so begin by taking an educated guess for the capacitors: $C_1 = 10$ μF, $C_2 = 2\,\mu F$, and $C_s = 1\,\mu F$.

The circuit operating-point extremes have already been calculated in Problem 16 for the extremes of IDSS and VP, as given by the specifications. At the high-current extreme of IDSS = 20 mA and VP = -3.5 volts, note that VGS = -1 volt. VGS is required to obtain values for the transistor capacitances, C-GS and C-GD. Figures 6 and 7 of the specifications for the Chip JN55 show that:

$$C_{rss} = 3.4\,\text{pF} = C_{gd}, \text{ and}$$
$$C_{gs} = C_{iss} - C_{gd} = 14 - 3.4 = 11.6\,\text{pF}.$$

RUNning the program shows that the low-frequency 3dB point is too high. Therefore, use branching code "1" and repeat the analysis. Increasing C_1 to 25 μF, C_2 to 4 μF, and C_s to 2 μF yields a more reasonable result for the low-frequency 3dB point. Note that capacitor C_1 controls, and that the 3dB points due to the other two are both less than 10% of that of the control. Therefore, elect to proceed, by using branching code "2".

The low-current extreme is given by IDSS = 5 mA and VP = − .5 volts, resulting in VGS = − 0.1 volts. (See the results of Problem 16.) IN-PUTing the new values for the transistor's capacitances and the new transconductance yields the final results. Although the common-gate amplifier differs somewhat from the common-source, it is again the low-current population which will have the minimum bandwidth, as the high-frequency 3dB point is 3 MHz less than that of the high-current extreme member of the transistor's population.

We conclude that our design meets the following specifications:

$$46 \text{ ohms} \le \text{Input Impedance} \le 79 \text{ ohms}$$

$$4.38 \le \text{Overall Voltage Gain} \le 7.46$$

$$48 \le \text{Low-Frequency 3dB Point} \le 64$$

$$46 \text{ MHz} \le \text{High-Frequency 3dB Point} \le 49 \text{ MHz}$$

Computer Results for Problem 17

```
A SIMPLIFIED LOW AND HIGH FREQUE
NCY ANALYSIS FOR THE COMMON BASE
 AMPLIFIER

R1=   1200000

R2=   51000

RD=   1000

R=   200

RS=   52

RL=   20000

C1=   .00001

C2=   .000002

CS=   .000001

C-GS=   1.16E-11

C-GD=   3.4E-12

GM AT OP.POINT=   .00762

THE LOW FREQUENCY 3 DB POINT IS
```

```
THE LARGER OF THE FOLLOWING THRE
E FREQUENCIES IN HZ:
F-C1=   121.2707989
F-C2=   3.789403592
F-CS=   3.253314437


THE HIGH FREQUENCY 3 DB POINT IS
 THE SMALLER OF THE FOLLOWING TW
O FREQUENCIES:
F-CGD=   49150793.65
F-CGD=   437000656.2


C1=   .000025

C2=   .000004

CS=   .000002

THE LOW FREQUENCY 3 DB POINT IS
THE LARGER OF THE FOLLOWING THRE
E FREQUENCIES IN HZ:
F-C1=   48.50831957
F-C2=   1.894701796
F-CS=   1.626657218


C-GS=   1.44E-11

C-GD=   3.6E-12

GM AT OP,POINT=   .01667

THE LOW FREQUENCY 3 DB POINT IS
THE LARGER OF THE FOLLOWING THRE
E FREQUENCIES IN HZ:
F-C1=   64.86407601
F-C2=   1.894701796
F-CS=   1.626657218


THE HIGH FREQUENCY 3 DB POINT IS
 THE SMALLER OF THE FOLLOWING TW
O FREQUENCIES:
F-CGD=   46420194.
F-CGD=   452052772.1


THIS COMPLETES THE ANALYSIS
```

PROBLEM 18 ———

Calculate the currents I_1, I_2, I_3, and I_4 in the following circuit, taken from a principal text * in circuit theory (there, the circuit was analyzed using a large computer).

Solution:

Use Program 16. Begin by writing the mesh equations using the Kirchoff Voltage Laws. The result may be written in the following form:

$$
\begin{bmatrix} 1 + j0 \\ 0 + j0 \\ 0 + j0 \\ 0 + j0 \end{bmatrix}
=
\begin{bmatrix}
0.1 + j2 & 0 + j1 & 0 + j0 & 0 + j0 \\
0 + j1 & 0.1 + j1 & 0 + j1 & 0 + j0 \\
0 + j1 & 0 + j1 & 0.1 + j1 & 0 + j1 \\
0 + j0 & 0 + j0 & 0 + j1 & 0.1 + j1
\end{bmatrix}
\mathbf{X}
\begin{bmatrix} \Phi_1 \\ \Phi_2 \\ \Phi_3 \\ \Phi_4 \end{bmatrix}
$$

INPUTing the impedance matrix and the voltage vector yields the following results:

```
SOLUTION OF THE MATRIX EQUATION
   V=Z*I BY WAY OF THE MODIFIED
   GAUSS-JORDAN ALGORITHM

N=  4

ENTER THE Z MATRIX SEPERATING TH
E REAL AND IMAGINARY PARTS BY A
COMMA:
R( 1 , 1 )= .1
X( 1 , 1 )= 2

R( 1 , 2 )= 0
X( 1 , 2 )= 1
```

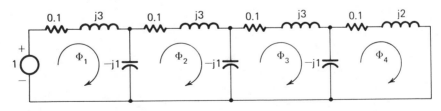

FIGURE 18.1. The four mesh circuit for problem 18.

———

* "Basic Circuit Theory", by Lawrence P. Huelsman, Prentice-Hall Inc., 1972.

```
R( 1 , 3 )= 0
X( 1 , 3 )= 0

R( 1 , 4 )= 0
X( 1 , 4 )= 0

R( 2 , 1 )= 0
X( 2 , 1 )= 1

R( 2 , 2 )= .1
X( 2 , 2 )= 1

R( 2 , 3 )= 0
X( 2 , 3 )= 1

R( 2 , 4 )= 0
X( 2 , 4 )= 0

R( 3 , 1 )= 0
X( 3 , 1 )= 0

R( 3 , 2 )= 0
X( 3 , 2 )= 1

R( 3 , 3 )= .1
X( 3 , 3 )= 1

R( 3 , 4 )= 0
X( 3 , 4 )= 1

R( 4 , 1 )= 0
X( 4 , 1 )= 0

R( 4 , 2 )= 0
X( 4 , 2 )= 0

R( 4 , 3 )= 0
X( 4 , 3 )= 1

R( 4 , 4 )= .1
X( 4 , 4 )= 1
```

ENTER THE V MATRIX

```
REAL V( 1 )= 1
IMG  V( 1 )= 0

REAL V( 2 )= 0
IMG  V( 2 )= 0
```

```
REAL V( 3 )= 0
IMG  V( 3 )= 0

REAL V( 4 )= 0
IMG  V( 4 )= 0


THE MESH CURRENTS ARE:
REAL I( 1 )= .0733578923
IMG  I( 1 )=-.4962849756

REAL I( 2 )=-.0970872871
IMG  I( 2 )=-.0000942595

REAL I( 3 )= .0237388207
IMG  I( 3 )= .4866705064

REAL I( 4 )= .0246814158
IMG  I( 4 )=-.4842023648
```

The results of the computations using the author's TI–99/4 micro-computer system agree in every respect with those in the reference (see page 450) which appear to have been obtained by means of a single-precision computation with 8 significant figures in the Mesh Current printout. The FORMAT statement calls for fourteen-digit precision, but only seven after the decimal point, and the actual currents are all less than 1.

The agreement calls for some caution as it is due to the fact that the micro-computer used for the sample program runs and all of the problems in this book uses a 16-bit processor, running double-precision, with a 13-digit floating point computation. Some micro-computers would not yield such accuracy from their normal single-precision format, particularly if they happen to use an 8-bit processor. Some caution is advisable when the user needs to invert a large matrix!

PROBLEM 19 _____

Repeat Problem 18 using node theory, and compare the solutions for the voltage across the j2 inductor in the 4th mesh of Figure 18.1.

Solution:

It is best to begin by redrawing the circuit. The first step is to convert the 1-volt source to its Norton equivalent. The resulting circuit is illustrated by Figure 19.1. In solving this problem, I have elected to use 7 nodes rather then 4, to test Program 17 with the case of a 7×7 Matrix.

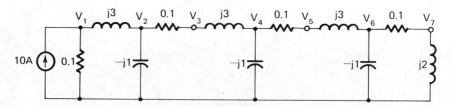

FIGURE 19.1. The seven node circuit for problem 19.

Using Kirchhoff's Current Law yields the following matrix equation for the node voltages V_1 through V_7:

$$
\begin{bmatrix} 10 \\ 0 \\ 0 \\ 0 \\ 0 \\ 0 \\ 0 \end{bmatrix}
=
\begin{bmatrix}
10-j1/3 & j1/3 & 0 & 0 & 0 & 0 & 0 \\
j1/3 & 10+j2/3 & -10 & 0 & 0 & 0 & 0 \\
0 & -10 & 10-j1/3 & j1/3 & 0 & 0 & 0 \\
0 & 0 & j1/3 & 10+j2/3 & -10 & 0 & 0 \\
0 & 0 & 0 & -10 & 10-j1/3 & -10 & 0 \\
0 & 0 & 0 & 0 & j1/3 & 10+j2/3 & -10 \\
0 & 0 & 0 & 0 & 0 & -10 & 10-j1/2
\end{bmatrix}
\times
\begin{bmatrix} V_1 \\ V_2 \\ V_3 \\ V_4 \\ V_5 \\ V_6 \\ V_7 \end{bmatrix}
$$

Using Program 17, enter the matrix equation, and obtain the following results:

```
SOLUTION OF THE MATRIX EQUATION
I=Y*V BY WAY OF THE MODIFIED
GAUSS-JORDAN ALGORITHM

N= 7

ENTER THE Y MATRIX SEPERATING TH
E REAL AND IMAGINARY PARTS BY A
COMMA:


THE NODE VOLATGES ARE:
REAL V( 1 )= .9926642108
IMG  I( 1 )= .0496284976

REAL V( 2 )=-.4961907163
IMG  I( 2 )=-.1704451795

REAL V( 3 )=-.4864819876
```

```
IMG   I( 3 )=-.1704357535

REAL  V( 4 )=-.4867647658
IMG   I( 4 )= .1208261079

REAL  V( 5 )=-.4891386479
IMG   I( 5 )= .0721590572

REAL  V( 6 )= .9708728713
IMG   I( 6 )= .000942595

REAL  V( 7 )= .9684047297
IMG   I( 7 )= .0493628315
```

In order to compare the result of the nodal analysis to that of the mesh analysis for the voltage across the last inductor, note from the results of Problem 18 that it is given by:

$$V_7 = I_4 * j2 = 0.9684047296 + j0.0493628316.$$

We have obtained the identical result for the first 9 significant figures, and a difference of 1 in the 10th. In view of the number of nodes, this is very good agreement.

PROBLEM 20 _____

Design active low-pass filters, using the Harris HA-2600, with 3dB frequencies of 10, 50, and 100 Hz.

Solution:

Use Program 18. Referring to the specification sheets for the HA-2600, note from the figure for the open-loop frequency response with C = 0 that the open-circuit gain is flat to a frequency of 100 Hz, with a gain of 150,000 (about 103dB). In view of this, meeting the highest frequency specification, 100 Hz, should be straightforward, because there will be no need to depend on the bandwidth improvement resulting from the feedback.

Using Program 18 yields the following results for the designs. For resistor R1′, one could use a 910-ohm standard; the capacitor is probably the most difficult to realize in practice.

Computer Results for Problem 20

```
DESIGN OF ACTIVE LOW PASS FILTER
S BY WAY OF THE BUTTERWORTH POLY
NOMIALS
```

```
SECOND ORDER FILTER DESIGN. REFE
R TO FIGURE 18B.

3 DB HIGH FREQUENCY LIMIT=  10
RESISTOR R=  1000
C=  .0000159155

THE GAIN AV=1.586

RESISTOR R1=  1600
R1'=  937.6

THIS COMPLETES YOUR DESIGN

SECOND ORDER FILTER DESIGN. REFE
R TO FIGURE 18B.

3 DB HIGH FREQUENCY LIMIT=  50
RESISTOR R=  1000
C=  .0000031831

THE GAIN AV=1.586

RESISTOR R1=  1600
R1'=  937.6

THIS COMPLETES YOUR DESIGN
SECOND ORDER FILTER DESIGN. REFE
R TO FIGURE 18B.

3 DB HIGH FREQUENCY LIMIT=  100
RESISTOR R=  1000
C=  1.59155E-06

THE GAIN AV=1.586

RESISTOR R1=  1600
R1'=  937.6

THIS COMPLETES YOUR DESIGN
```

PROBLEM 21 _____

given the following information for a satellite communications ground receiving system:

Received power at the antenna output $= -100\text{dBW}$, a receiver front end with a bandwidth of 36 MHz, and the specifications shown in Figure 21.1.

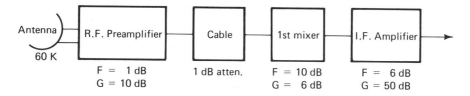

FIGURE 21.1. Block diagram of the first four stages of the communications receiver for a satellite ground station.

calculate the signal-to-noise ratio at the output of the IF amplifier.

Solution:

Begin by using Program 19 to calculate the noise figure of the receiving system.

```
NOISE FIGURE OF A COMMUNICATIONS
  RECEIVER

NUMBER OF STAGES=  4

F1=  1
G1=  10

F2=  1
G2= -1

F3=  10
G3=  6

F4=  6

NOISE FIGURE=  2.512120588

NOISE FIGURE IN DB=  4.000404828

EFFECTIVE NOISE TEMP. IN DEG. KE
LVIN=  438.5149706
```

You must calculate the signal-to-noise ratio at the output of the receiver's IF amplifier. The available gain of the receiver does not enter the calculation, because both signal and noise are multiplied by the same gain. However, you must use the effective temperature both of the antenna and of the receiver.* Consequently, write:

$$P_N = k * T_{eff} * B$$

---

*"Principles of Communications", by R.E. Ziemer and W.H. Tranter. Houghton Mifflin Company, 1976.

where

$$k = 1.38 \times 10^{-23} \text{ watt-seconds}/°K \text{ (Boltzman's constant)};$$
$$T_{eff} = T_{Ant.} + T_{eff(Amp.)} = 60 + 438.5 = 498.5°K; \text{ and}$$
$$B = 36 \text{ MHz (bandwidth in mega Hz)}.$$

The calculation shows that the noise power is -126dBW, or 126dB below 1 watt. The received power at the antenna output is -100dBW, so the signal-to-noise ratio at the output of the IF amplifier is 26dB.

PROBLEM 22 _____

Design a 1-watt power amplifier, based on the TIP29 transistor, for use in a telephone amplifier for those with hearing difficulties. The amplifier will work into a small 8-ohm loudspeaker. A 30-volt power supply is available.

Solution:

Use Program 20 for the design; then analyze the results with Program 21.

For good voice fidelity, target the DC power input at 3 watts. For RS, use 5.1K ohms, on the assumption that a voltage amplifier of this output impedance at mid-band will drive the power amplifier. The specifications for Beta AC and DC range from about 40 to 200; design to the average value of 120, at a current of 0.15 amps.

```
POWER AMP DESIGN

TARGET VCE=   20
TARGET IC=   .15
TARGET ZO=   8

VCC=   30
RS=   5100
AMBIENT TEMP=   25
TOTAL THERMAL RESISTANCE-JUNCTIO
N TO AMBIENT=   6.67
BETA DC=   120
BETA AC=   120
INITIAL GAMMA=   10

DESIGN GAMMA=   4
VCE=   20.0686771
IC=   .15
VBE=   .7
TJ=   44.94081667
DC POWER IN=   2.989627686
GM=   5.435131614
RPIE=   22.07858218
```

```
R1=   3860
R2=   2853.333333
RE=   66.66666667

ZO=   9.02725538
ZIN=  615.1263812
AVI=  .9775649031
AP=   65.11781602
```

Note that a heat sink is required at this power level; a smaller one could safely be used.

The power gain of the circuit will be about 65, necessitating a driver capable of delivering about 16 mW (no problem). One watt across 8 ohms requires about 2.8 volts RMS of drive. The voltage gain, AVI, is about .97, necessitating a driving voltage of about 2.9 volts at the input of the power amplifier. This information is required to design the driving voltage amplifier, a task that is left to the interested user, as it depends on the way in which the telephone signal is sampled.

The results of the computer-assisted analysis are for a 5%-resistor realization, with the capacitors selected to yield a low-frequency 3dB point of about 125 Hz (adequate for voice reproduction).

```
A SIMPLIFIED MID-BAND, LOW AND
HIGH FREQUENCY ANALYSIS FOR THE
EMITTER FOLLOWER AMPLIFIER

ALL VOLTAGE IN VOLTS, ALL CURENT
 IN AMPS, ALL RESISTANCE IN OHMS

R1=   3900
R2=   2700
RE=   68
RL=   8
RS=   5100
VCC=   30
BETA DC=   120
BETA AC=   120
AMBIENT TEMP=   25
THERMAL RESISTANCE RJA=   6.67
V-BE=   .7

VCE=   20.38694035
IC=   .1413685243
PD(VCE*IC)=   2.882071672
JUNCTION TEMP=   44.22341805
GM=   5.169463503
RPIE=   23.21324059
ZIN=   571.0248868
```

```
Z-OUT=  8.896350496

AVI=  .97389772
AVS=  .0980633742
AI=  69.51497942
AP=  67.70047996

A=  1
LOW FREQ ANALYSIS

C1=  .000003
C2=  .000075

THE LOW FREQ 3 DB POINT IS THE L
OWER OF THE FOLLOWING TWO FREQ.
FC1=  9.354860673
FC2=  125.5931515


    THIS COMPLETES YOUR ANALYSIS
```

PROBLEM 23 _____

Design an oscillator using the 2N5359 FET transistor for a frequency at the high end of the broadcast band, say 1.6 MHz. Use a 100-picofarad capacitor and a 7- to 25-picofarad trimmer for fine tuning. A 6-volt battery is available as a power source.

> **Notice:** *Radio transmitters are regulated by the Federal Communications Commission; the user should take care not to break the laws.*

Solution:

Use Program 22. As the input for the capacitor C, use 100 picofarads plus one-half the maximum value of the trimmer, 25 picofarads. For the resistor RD assume 100K ohms, for reasons explained in our example for Program 22. As input for the mutual inductance M, use 5 microhenries. Note that L1 should be about 81 microhenries. From the specification sheets for the transistor, the average values for IDSS and VP are:

$$IDSS = 1.2 \text{ mA}, \qquad VP = -1.2 \text{ volts.}$$

The maximum transconductance, g_m, will once again by 2.E − 3 mhos, for the average transistor. Since the value of g_m required for oscillation is predicted by the computer as 0.375 milli-mohs, the design will be feasible.

INPUTing IDSS and VP yields a value for the maximum absolute value of VGS of 0.975 volts. Design for VGS = − .6 V.

```
FET TUNED DRAIN OSCILLATOR
 DESIGN

OCS. FREQ. IN HZ=  1600000
CAPACITOR C=  1.12E-10
L1=  .0000883451
Q OF COIL L1=  100

RD=  100000
MUTUAL INDUCTANCE M=  .000005

GM MIN FOR OSCILATION=
 .0003756338

IDSS=  .0012
VP= -1.2

ABS(VGS MAX)=  .9746197194

VGS= -.6
THE MINIMUM POWER SUPPLY VOLTAGE
 VDD=  1.6


VDD=  6
VDS=  5.4
ID=  .0003
R=  2000
L1=  .0000883451
M=  .000005
C=  1.12E-10
F=  1600000
```

The schematic is similar to that of Figure 22; capacitor C must be replaced by the parallel combination of C and the trimmer. Should the user wish to wind the coils, helpful information can be found in the Radio Amateur's Handbook, published yearly by the American Radio Relay League.

PROBLEM 24 _____

Calculate the transfer characteristics and the drain current of a type CD4007A C MOS Inverter loaded with a 10K-ohm resistor, as illustrated on page 210.

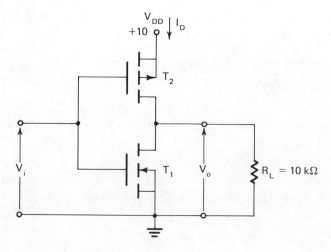

FIGURE 24.1 An integrated circuit CMOS inverter loaded by a 10K ohm resistor.

Solution:

Use Program 23. From our analysis of the example, $k_n = 1.875 \times 10^{-4}$ amps/volt^2 and $k_p = 2.5 \times 10^{-4}$ amps/volt^2. Take RL = 10K Ohms. The results of the computer run follow. (A hard copy of the PRINT statements is very useful.)

The results have been plotted as Figures 24.2 and 24.3. A comparison of these results to those of the example in Program 23 indicates that while the transfer characteristic is not changed significantly, the drain current is considerably different, because of the current of about 0.97 mA which will be drawn when the output is in Logic 1. This is due to the 10K-ohm load resistor. Note also that T_2 draws current in Logic 1, resulting in a somewhat lower Logic 1 output voltage, 9.735 volts rather than the previous 10 volts.

Computer Results for Problem 24

```
TRANSFER CHARACTERISTIC OF AN IC
 CMOS INVERTER

VSS=  10
VT-N=  2.5
VT-P=  2.5
KN=  .0001875
KP=  .00025
RL=  10000
VIMAX= 10 VIMIN= 0
NO OF STEPS=  50
```

```
VI=   0
R(#1)
VO=   9.735
I( 0 )= .000973

VI=   .2
R(#1)
VO=   9.728
I( 1 )= .000972

VI=   .4
R(#1)
VO=   9.72
I( 2 )= .000972

VI=   .6
R(#1)
VO=   9.712
I( 3 )= .000971

VI=   .8
R(#1)
VO=   9.703
I( 4 )= .00097

VI=   1
R(#1)
VO=   9.694
I( 5 )= .000969

VI=   1.2
R(#1)
VO=   9.684
I( 6 )= .000968

VI=   1.4
R(#1)
VO=   9.674
I( 7 )= .000967

VI=   1.6
R(#1)
VO=   9.662
I( 8 )= .000966

VI=   1.8
R(#1)
VO=   9.65
I( 9 )= .000965
```

```
VI=   2
R(#1)
VO=   9.637
I( 10 )= .000963

VI=   2.2
R(#1)
VO=   9.623
I( 11 )= .000962

VI=   2.4
R(#1)
VO=   9.608
I( 12 )= .00096

VI=   2.6
R(#2)
VO=   9.59
I( 13 )= .00096

VI=   2.8
R(#2)
VO=   9.565
I( 14 )= .000973

VI=   3
R(#2)
VO=   9.531
I( 15 )= .000999

VI=   3.2
R(#2)
VO=   9.485
I( 16 )= .00104

VI=   3.4
R(#2)
VO=   9.425
I( 17 )= .001094

VI=   3.6
R(#2)
VO=   9.35
I( 18 )= .001161

VI=   3.8
R(#2)
VO=   9.253
I( 19 )= .001242
```

```
VI=   4
R(#2)
VO=   9.128
I( 20 )= .001334


VI=   4.2
R(#2)
VO=   8.966
I( 21 )= .001438


VI=   4.4
R(#2)
VO=   8.745
I( 22 )= .001551


VI=   4.6
R(#2)
VO=   8.417
I( 23 )= .001668


VI=   4.8
R(#2)
VO=   7.765
I( 24 )= .001768


VI=   5
R(#2)
R(#3)
VO=   3.906
I( 25 )= .001562


VI=   5.2
R(#3)
VO= -.444
I( 26 )= .001322


R(#4)
VO=   1.644
I( 26 )= .001322


VI=   5.4
R(#4)
VO=   1.13
I( 27 )= .001102


VI=   5.6
R(#4)
VO=   .813
```

```
VI=  5.8
R(#4)
VO=  .588
I( 29 )= .000722

VI=  6
R(#4)
VO=  .421
I( 30 )= .000562

VI=  6.2
R(#4)
VO=  .295
I( 31 )= .000422

VI=  6.4
R(#4)
VO=  .198
I( 32 )= .000302

VI=  6.6
R(#4)
VO=  .125
I( 33 )= .000202

VI=  6.8
R(#4)
VO=  .072
I( 34 )= .000122

VI=  7
R(#4)
VO=  .035
I( 35 )= .000062

VI=  7.2
R(#4)
VO=  .012
I( 36 )= .000022

VI=  7.4
R(#4)
VO=  .001
I( 37 )= .000002

VI=  7.6
R(#4)

AVERAGE POWER=  .0092450496
```

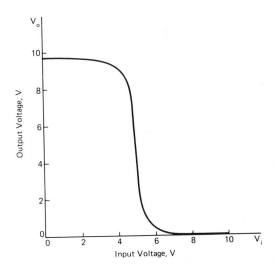

FIGURE 24.2 Transfer Characteristics for a Type CD4007A Inverter with a 10K-ohm Load.

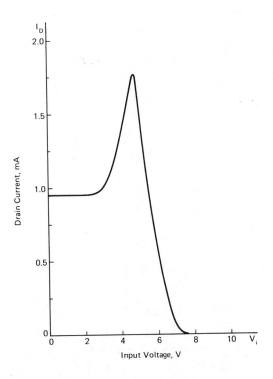

FIGURE 24.3 Drain Current as a Function of the Input Voltage

PROBLEM 25 _____

Design an unregulated power supply, using a capacitor filter, to be used in conjunction with an integrated-circuit regulator to yield a regulated voltage of about 9 volts with a maximum load current of about 1.5 amps.

Solution:

The exact solution will depend on the regulator used. Assume that the unregulated part of the power supply must produce a loaded voltage of about 15 volts at a load current of about 2 amperes, thereby giving our hypothetical regulator a voltage drop of at most 6 volts and an excess current of about 0.5 amps. This should be more than sufficient. The regulator will also decrease the ripple voltage, so assume that an unregulated ripple factor of 0.01 will suffice.

Use Program 24.

Computer Results for Problem 25

```
DESIGN OF AN UNREGULATED FULL
  WAVE BRIDGE RECTIFIER

ALL VOLTAGES IN VOLTS.
ALL CURRENTS IN AMPS.

TARGET DC OUTPUT VOLTAGE=  15
TARGET DC OUTPUT CURRENT=  2
TARGET RIPPLE FACTOR=  .01
DIODE VOLTAGE DROP=  1

CAPACITOR FILTER DESIGN

REQUIRED FILTER CAPACITOR IN MIC
ROFARADS=  32000

% REGULATION=  15.08333333

PEAK TRANSFORMER VOLTAGE=
  17.2625

SECONDARY RMS TRANSFORMER VOLTAG
E=  12.20643081

PEAK-TO-PEAK RIPPLE VOLTAGE=
  .5196152423

PEAK DIODE FORWARD CURRENT=
  23.01666667
```

```
PEAK REVERSE DIODE VOLTAGE=
   34.525
```

A 12-volt transformer may be used, such as the Stancor Type P–8657. A user wishing to design a regulator using discrete components is referred to the texts noted below.

PROBLEM 26 _____

Design a full-wave bridge rectifier power supply with a pi filter for an output of 600 volts at a load current of 0.15 amperes. Target the ripple factor at 0.0001 so as to obtain a peak-to-peak ripple voltage of less than 0.1 volts.

Solution:

Use Program 24.

Computer Results for Problem 26

```
DESIGN OF AN UNREGULATED FULL
   WAVE BRIDGE RECTIFIER

ALL VOLTAGES IN VOLTS.
ALL CURRENTS IN AMPS.

TARGET DC OUTPUT VOLTAGE=   600
TARGET DC OUTPUT CURRENT=   .15
TARGET RIPPLE FACTOR=   .0001
DIODE VOLTAGE DROP=   8

PIE FILTER DESIGN

CHOKE INDUCTANCE=   8
CHOKE DC RESISTANCE=   145

REQUIRED FILTER CAPACITORS IN MI
CROFARADS=   32.11308145

% REGULATION=   9.561362232

PEAK TRANSFORMER VOLTAGE=
   657.3681734
```

1. M. M. Cirovic, "Basic Electronics", 2nd Edition. Reston Publishing Company, Reston, Virginia. See Chapter 19.

2. J. Millman and C. C. Halkias, "Integrated Electronics: Analog and Digital Circuits and Systems", McGraw-Hill Book Company, 1972. See Sections 18–9 and 18–10.

```
SECONDARY RMS TRANSFORMER VOLTAG
E=  464.8294931

PEAK-TO-PEAK RIPPLE VOLTAGE=
  .0848528137

PEAK DIODE FORWARD CURRENT=
  1.643420433

PEAK REVERSE DIODE VOLTAGE=
  1314.736347
```

Comments:

Refer to the schematic in Program 24. Since electrolytic capacitors with a working voltage of more than 600 volts are difficult to obtain, high-voltage power supplies are usually designed with capacitors in series, each capacitor bypassed by a resistor for safety. For example, the two 32-micro farad capacitors could be realized by using two (for each) 80-micro farad capacitors in series. Each capacitor could then have a working DC voltage of 350 volts, a standard size, although 450 volts would give a greater safety factor. Since the percentage of regulation is just under 10%, we may expect a no-load voltage of about 660 volts. A possible configuration is illustrated by Figure 26.1 below.

Caution: Extreme caution is advised in reducing any very high-voltage power supply to practice. The inexperienced user should take great care, because very high-voltage power supplies, used improperly, can cause fatalities. If in doubt, please *consult an engineer*.

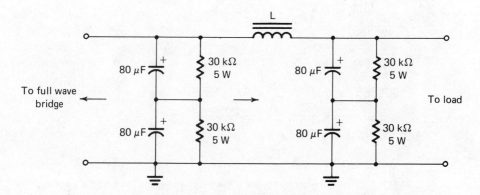

FIGURE 26.1. A possible filter realization for the power supply.

PROBLEM 27 _____

Design a three-stage voltage amplifier with a mid-band overall voltage gain of at least 5,000, an input impedance of at least 5K ohms, and an output impedance of about 52 ohms. A 15-volt power supply is available. Use type 2N5089 transistors, so as to easily meet the high input-impedance specification. After the amplifier is designed, analyze your results for the maximum and minimum values of all design targets.

Solution:

The problems of multi-stage design have been discussed in Program 25. Begin by designing the 3rd stage. Since a mid-band output impedance of 52 ohms is required, an emitter follower is the logical choice. Such an amplifier has already been designed and analyzed as the example for Program 8. Use it as the initial design for stage 3. From the example, note that the mid-band input impedance will vary from about 13K to 16K ohms. Use the average value, 14.5K ohms, as the load resistor for stage 2.

    Because R_c of stage 1 will be the input or source impedance for stage 2, first design stage 1 using Program 5, for an overall voltage gain of at least 100.

Design of Stage 1

```
COMMON EMMITER, MID-  BAND VOLTA
GE AMPLIFIER DESIGN
ALL R IN OHMS
ALL I IN AMPS
ALL V IN VOLTS

ENTER MINIMUM ALLOWABLE VOLTAGE
GAIN AT MIDBAND
AV MIN=  100

ENTER MINIMUM ALLOWABLE INPUT IM
PEDANCE
ZIN MIN=  5000

ENTER SOURCE IMPEDANCE
ZS=  500

ENTER POWER SUPPLY VOLTAGE
VCC=  15

ENTER JUNCTION TEMP. IN DEG. C
TJ=  25
```

```
ENTER LOAD RESISTOR
RL=  5000

ENTER COLLECTOR RESISTOR
RC=  7500

THE LOWER BOUND ON GM IS GM=
 .0366666667

THE LOWER BOUND ON THE COLLECTOR
 CURRENT IS, IC= .000941954

CHOOSE A DESIGN VALUE FOR THE CO
LLECTOR CURRENT
IC=  .001

ENTER MINIMUM TRANSISTOR SMALL S
IGNAL BETA AT THE OPERATING POIN
T
MIN BETA AC=  450

THE SMALL SIGNAL TRANSISTOR INPU
T IMPEDANCE R-PIE=  11560.34483


ENTER MINIMUM D.C.BETA AT THE OP
ERATION POINT
MIN BETA DC=  450

DESIGNER MUST ENTER PARAMETER GA
MMA
GAMMA=  50

ZIN= 8214.74898


R1= 95400.

R2= 40408.16327

RE= 3750

RC= 7500

RL= 5000

AVS= 103.6888601

THIS COMPLETES YOUR COMPUTER DES
IGN. TO REPEAT FOR DIFFERENT GAM
```

```
MA, ENTER 1, TO END ENTER 0.
X=  1

DESIGNER MUST ENTER PARAMETER GA
MMA
GAMMA=  40

ZIN= 8727.121116


R1= 119250.

R2= 50769.23077

RE= 3750

RC= 7500

RL= 5000

AVS= 104.0393109

THIS COMPLETES YOUR COMPUTER DES
IGN. TO REPEAT FOR DIFFERENT GAM
MA, ENTER 1, TO END ENTER 0.
X=  0
```

Next, analyze the design, using the results for GAMMA = 40 and a realization with 5% resistors, with Program 6. Note that this program has been modified in order to calculate and print AVI (defined in the Glossary). This is given by the equations:

$$AVI\ MAX = GMM * RL1 \qquad \text{and} \qquad AVI\ MIN = GMN * RL1.$$

If the user is planning many multi-stage designs, this addition should prove useful.

Analysis of Stage 1

```
ANALYSIS OF A COMMON EMITTER BJT
 AMPLIFIER FOR RANGE OF D.C. AND
 A.C. OP.

R1=  120000

R2=  51000

RE=  3600

RC=  7500
```

```
RL=   5000

RS=   500

VCC=  15

BETA DC MAX=   1350
BETA DC MIN=   450

BETA AC MAX=   1800
BETA AC MIN=   450

TEMP DEG. C=   25

I-CEO=   0

ICMAX=   .0010535953
ICMIN=   .0010369223

VCE MAX=   3.498439822
VCE MIN=   3.307899764

ZIN MAX=   19713.80984
ZIN MIN=   8500.679197

AVI MAX=   123.0373026
AVI MIN=   121.0902513

AVS MAX=   119.9939054
AVS MIN=   114.3635228
```

Next, design the 2nd stage. The source impedance is R_L of the 1st stage, 7.5K ohms. The load impedance is about 14.5K ohms. The gain of importance here is AVI, rather then AVS, as explained in Program 25. The printout of the design following is from Program 5.

Design of Stage 2

```
COMMON EMMITER, MID-  BAND VOLTA
GE AMPLIFIER DESIGN
ALL R IN OHMS
ALL I IN AMPS
ALL V IN VOLTS

ENTER MINIMUM ALLOWABLE VOLTAGE
GAIN AT MIDBAND
AV MIN=   50

ENTER MINIMUM ALLOWABLE INPUT IM
```

```
PEDANCE
ZIN MIN=  5000

ENTER SOURCE IMPEDANCE
ZS=  7500

ENTER POWER SUPPLY VOLTAGE
VCC=  15

ENTER JUNCTION TEMP. IN DEG. C
TJ=  25

ENTER LOAD RESISTOR
RL=  14500

ENTER COLLECTOR RESISTOR
RC=  3900

THE LOWER BOUND ON GM IS GM=
 .0406719717

THE LOWER BOUND ON THE COLLECTOR
 CURRENT IS; IC= .0010448489

CHOOSE A DESIGN VALUE FOR THE CO
LLECTOR CURRENT
IC=  .0011


ENTER MINIMUM TRANSISTOR SMALL S
IGNAL BETA AT THE OPERATING POIN
T
MIN BETA AC=  450

THE SMALL SIGNAL TRANSISTOR INPU
T IMPEDANCE R-PIE=  10509.40439


ENTER MINIMUM D.C.BETA AT THE OP
ERATION POINT
MIN BETA DC=  450

DESIGNER MUST ENTER PARAMETER GA
MMA
GAMMA=  50

ZIN= 7770.792234


R1= 73595.45455
```

```
R2= 50134.50835

RE= 4868.181818

RC= 3900

RL= 14500

AVI=  125
AVS= 63.60829316


THIS COMPLETES YOUR COMPUTER DES
IGN. TO REPEAT FOR DIFFERENT GAM
MA, ENTER 1, TO END ENTER 0.
X=  0
```

Next, analyze the design for stage 2, using Program 6.

Analysis of Stage 2

```
ANALYSIS OF A COMMON EMITTER BJT
 AMPLIFIER FOR RANGE OF D.C. AND
 A.C. OP.

R1=  75000

R2=  51000

RE=  4700

RC=  3900

RL=  14500

RS=  7500

VCC=  15

BETA DC MAX=  1350
BETA DC MIN=  450

BETA AC MAX=  1800
BETA AC MIN=  450

TEMP DEG. C=  25

I-CEO=  0

ICMAX=  .0011471572
ICMIN=  .0011346875
```

```
VCE MAX=   5.253512831
VCE MIN=   5.138438818

ZIN MAX=   17316.25728
ZIN MIN=   7628.079123

AVI MAX=   137.2396025
AVI MIN=   135.7477884

AVS MAX=   95.76287994
AVS MIN=   68.44853617
```

Next, redo the analysis for the emitter-follower amplifier of stage 3 using the correct source resistance, the R_c of stage 2, 3.9K ohms. The results for two values of RE follow:

Run # 1

```
ANALYSIS OF AN EMITTER FOLLOWER
BJT AMPLIFIER FOR RANGE OF D.C.
AND A.C. OP

R1=   56000

R2=   27000

RE=   8200

RL=   52

RS=   3900

VCC=   15

BETA DC MAX=   1350
BETA DC MIN=   450

BETA AC MAX=   1800
BETA AC MIN=   450

AMBIENT TEMP IN DEG C =   25

THERMAL RESISTANCE, JUNCTION TO
AMBIENT, IN DEG. C/WATT=   357

I-CEO=   0

ICMAX=   .000514567
ICMIN=   .0005121285
```

```
VCE MAX=   10.80985743
VCE MIN=   10.7836742

T-J MAX DEG C = 26.98096528
T-J MIN DEG C = 26.97636502

ZIN MAX=   16571.91342
ZIN MIN=   13051.33207

AVI MAX=   .5070830637
AVI MIN=   .5063156159

AVS MAX=   .4104812508
AVS MIN=   .3898273721

MAX CURRENT GAIN=   161.6026275
MIN CURRENT GAIN=   127.0787161

MAX POWER GAIN=   81.94595543
MIN POWER GAIN=   64.34193839

MAX OUTPUT Z=   57.10535333
MIN OUTPUT Z=   51.68455849
```

Run # 2

```
ANALYSIS OF AN EMITTER FOLLOWER
BJT AMPLIFIER FOR RANGE OF D.C.
AND A.C. OP

R1=   56000

R2=   27000

RE=   7500

RL=   52

RS=   3900

VCC=   15

BETA DC MAX=   1350
BETA DC MIN=   450

BETA AC MAX=   1800
BETA AC MIN=   450

AMBIENT TEMP IN DEG C =   25
```

```
THERMAL RESISTANCE, JUNCTION TO
AMBIENT, IN DEG. C/WATT=   357

I-CEO=   0

ICMAX=   .000562507
ICMIN=   .0005596712

VCE MAX=   10.81177352
VCE MIN=   10.78432015

T-J MAX DEG C = 27.16565328
T-J MIN DEG C = 27.16022048

ZIN MAX=   16506.51922
ZIN MIN=   12891.2414

AVI MAX=   .5290195066
AVI MIN=   .5281793256

AVS MAX=   .4279157341
AVS MIN=   .4055023108

MAX CURRENT GAIN=   167.9282818
MIN CURRENT GAIN=   130.9401382

MAX POWER GAIN=   88.83733675
MIN POWER GAIN=   69.1598739

MAX OUTPUT Z=   52.87869103
MIN OUTPUT Z=   47.4576878
```

Note that both designs meet the specification on the output impedance at mid-band.

Having analyzed all three stages, calculate the overall mid-band voltage gain. This is given by:

$$A_{Vs} \text{ (Overall)} = A_{Vs_1} * A_{Vi_2} * A_{Vi_3}.$$

For AVS MAX (taking RE of the 3rd stage as 7.5K ohms):

$$A_{Vs\,Max} = 120 * 137 * 0.529 = 8,696.$$

For AVS MIN, we obtain:

$$A_{Vs\,Min} = 114 * 135 * 0.528 = 8,126.$$

This design easily meets the original specification of at least 5,000. This is really a margin of safety; the gain may easily be reduced, and most circuits are reduced to practice with a potentiometer. Increasing the gain of a faulty design, however, is difficult.

The reduction for analysis employs 5% resistors. A schematic of the three-stage amplifier follows. The user may wish to specify the coupling capacitors for a target overall low-frequency 3dB point. For details, please refer to Program 25.

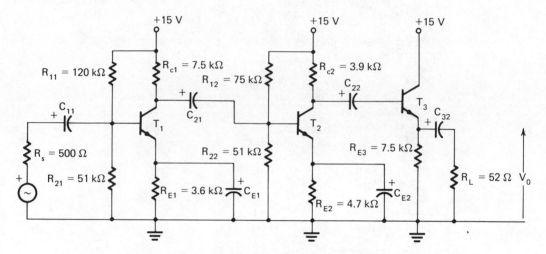

FIGURE 27.1. A possible design for the three stage amplifier.

Note:

The stages may require some decoupling if tying all three 15-volt terminals together results in low-frequency oscillations, by way of feedback through the power supply. If decoupling is required, try a 1K ohm resistor and a 10-micro farad capacitor attached to each 15-volt terminal, and connect the resistor to the power supply. This is common practice in high-gain amplifiers.

PART III
APPENDICES

APPENDIX A

Glossary of Terms Used in the Computer Listings

AI	Current Gain (I_o/I_i)
AP	Power Gain ($V_o \cdot I_o/V_i \cdot I_i$)
AVI	Circuit Voltage Gain (V_o/V_i)
AVS	Overall Circuit Voltage Gain (V_o/V_s)
BETA AC	Small-signal common-emitter short-circuit current gain
BETA DC	DC common-emitter short-circuit current gain
BJT	Bipolar Junction Transistor
C	Any capacitor in Farads
C–GD	The Gate-to-Drain capacitance of a Field Effect Transistor
C–GS	The Gate-to-Source capacitance of a Field Effect Transistor
C–ISS	Common-source short-circuit input capacitance of a Field Effect Transistor
C–MU	Base-to-Collector capacitance of a BJT
C–PI	Base-to-Emitter capacitance of a BJT
C–RSS	Short-circuit reverse transfer capacitance of a Field Effect Transistor
F	Noise Figure
F–DB	Noise Figure in decibels
FET	Field Effect Transistor
FH	The high-frequency 3dB point in Hz

FL	The low-frequency 3dB point in Hz
FT	Gain-Bandwidth Produce—the frequency in Hz at which the current gain is equal to unity
HFE	Another common notation for the DC BETA
HIE	Input impedance for the BJT Hybrid Model
HOE	Output admittance for the BJT Hybrid Model
HRE	Voltage feedback ratio in the BJT Hybrid Model
I–CBO	Reverse saturation current, BJT common-base configuration
I–CEO	Reverse saturation current, BJT common-emitter configuration
IB	Base current (DC or AC)
IC	Collector current (DC or AC)
ID	Drain current (DC or AC)
IDSS	Drain saturation current
KN	Parameter in an N-channel MOSFET's drain-current model equation
KP	Parameter in a P-channel MOSFET's drain-current model equation
MOSFET	Metal Oxide Semiconductor Field Effect Transistor
OP–AMP	Operational Amplifier
R	Any resistor in Ohms
RBC	Small-signal BJT base-collector resistance
RC	Small-signal BJT collector-emitter resistance
r	Ripple factor (see the Power-Supply Design Program)
RMS	Root-Mean-Squared value of a voltage or current
R–PI	Small-signal input resistance of the BJT Hybrid-Pi Model
RX	Transistor input lead resistance, Hybrid-Pi BJT Model (base to virtual base)
S/N	Signal-to-Noise Ratio
TA	Ambient temperature, usually in degrees Centigrade
TJ	Junction temperature of a BJT, usually in degrees Centigrade
VBE	Base-emitter voltage
VCC	Power supply voltage for a BJT
VCE	Collector-emitter voltage
VDD	Power supply voltage for a FET
VDS	Drain-to-Source voltage
VGS	Gate-to-Source voltage
VI	Circuit input voltage
VO	DC voltage between the base and emitter terminals (a parameter of the piece-wise linear model used by these programs to model the DC operating point of a BJT)
VP	Pinch-off voltage for a FET
VS	Source voltage for the small-signal AC analysis
VSD	Source-to-Drain voltage

VSG	Source-to-Gate voltage
VT	Thermal potential of the junction of a BJT

$$VT = \frac{T_J \, {}^\circ K}{11{,}600}, \text{ in Volts}$$

Y	Admittance in Mhos
Z	Impedance in Ohms
ZIN	Input impedance
ZO	Output impedance
ZS	Source impedance, usually assumed to be real

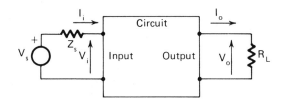

Two-Port Circuit Used to Define the Input and Output Variables Used in the Programs.

APPENDIX B

Transistor Specification Sheets

HA-2600/2602/2605

Wide Band, High Impedance Operational Amplifier

FEATURES

- WIDE BANDWIDTH   12MHz
- HIGH INPUT IMPEDANCE   500MΩ
- LOW INPUT BIAS CURRENT   1nA
- LOW INPUT OFFSET CURRENT   1nA
- LOW INPUT OFFSET VOLTAGE   0.5mV
- HIGH GAIN   150K V/V
- HIGH SLEW RATE   7V/μs
- OUTPUT SHORT CIRCUIT PROTECTION

APPLICATIONS

- VIDEO AMPLIFIER
- PULSE AMPLIFIER
- AUDIO AMPLIFIERS AND FILTERS
- HIGH-Q ACTIVE FILTERS
- HIGH-SPEED COMPARATORS
- LOW DISTORTION OSCILLATORS

DESCRIPTION

HA-2600/2602/2605 are internally compensated bipolar operational amplifiers that feature very high input impedance (500 MΩ, HA-2600) coupled with wideband AC performance. The high resistance of the input stage is complemented by low offset voltage (0.5mV, HA-2600) and low bias and offset current (1nA, HA-2600) to facilitate accurate signal processing. Input offset can be reduced further by means of an external nulling potentiometer. 12MHz unity gain-bandwidth product, 7V/μs slew rate and 150,000V/V open-loop gain enables HA-2600/2602/2605 to perform high-gain amplification of fast, wideband signals. These dynamic characterisitics, coupled with fast settling times, make these amplifiers ideally suited to pulse amplification designs as well as high frequency (e.g. video) applications. The frequency response of the amplifier can be tailored to exact design requirements by means of an external bandwidth control capacitor.

In addition to its application in pulse and video amplifier designs, HA-2600/2602/2605 is particularly suited to other high performance designs such as high-gain low distortion audio amplifiers, high-Q and wideband active filters and high-speed comparators.

HA-2600 and HA-2602 are guaranteed over -55ºC to +125ºC. HA-2605 is specified from 0ºC to +75ºC. All devices are available in TO-99 cans, and HA-2600/2602 are available in 10 lead flat packages.

PINOUT

TO-99

BANDWIDTH CONTROL

Package Code 2A, 1N, LA

OFFSET ADJ. 1
IN- 2
IN+ 3
4

8
7 V+
6 OUT
5 OFFSET ADJ.

2A

Case Connected to V–

V–

TOP VIEWS

IN+ 1   8 BANDWIDTH CONTROL
IN- 2   7 V+
OFFSET ADJ. 3 6 OUT
V- 4   5 OFFSET ADJ

1N

CAUTION: These devices are sensitive to electrostatic discharge. Users should follow IC Handling Procedures specified on pg. 1-4.

SCHEMATIC

SPECIFICATIONS

ABSOLUTE MAXIMUM RATINGS

Voltage Between V+ and V− Terminals	45.0V
Differential Input Voltage	±12.0V
Peak Output Current	Full Short Circuit Protection
Internal Power Dissipation	300mW
Operating Temperature Range − HA-2600/HA-2602	$-55^{\circ}C \leq T_A \leq +125^{\circ}C$
HA-2605	$0^{\circ} \leq T_A \leq +75^{\circ}C$
Storage Temperature Range	$-65^{\circ}C \leq T_A \leq +150^{\circ}C$

ELECTRICAL CHARACTERISTICS   V+ = +15VDC, V− = −15VDC

PARAMETER	TEMP.	HA-2600 −55°C to +125°C LIMITS MIN.	TYP.	MAX.	HA-2602 −55°C to +125°C LIMITS MIN.	TYP.	MAX.	HA-2605 0°C to +75°C LIMITS MIN.	TYP.	MAX.	UNITS
INPUT CHARACTERISTICS											
* Offset Voltage	+25°C		0.5	4		3	5		3	5	mV
	Full		2	6			7			7	mV
. Offset Voltage Average Drift	Full		5								µV/°C
* Bias Current	+25°C		1	10		15	25		5	25	nA
	Full		10	30			60			40	nA
* Offset Current	+25°C		1	10		5	25		5	25	nA
	Full		5	30			60			40	nA
Input Resistance (Note 10)	+25°C	100	500		40	300		40	300		MΩ
Common Mode Range	Full	±11.0			±11.0			±11.0			V
TRANSFER CHARACTERISTICS											
* Large Signal Voltage Gain (Notes 1, 4)	+25°C	100K	150K		80K	150K		80K	150K		V/V
	Full	70K			60K			70K			V/V
* Common Mode Rejection Ratio (Note 2)	Full	80	100		74	100		74	100		dB
Unity Gain Bandwidth (Note 3)	+25°C		12			12			12		MHz
OUTPUT CHARACTERISTICS											
Output Voltage Swing (Note 1)	Full	±10.0	±12.0		±10.0	±12.0		±10.0	±12.0		V
* Output Current (Note 4)	+25°C	±15	±22		±10	±18		±10	±18		mA
Full Power Bandwidth (Note 4 & 11)	+25°C	50	75		50	75		50	75		kHz
TRANSIENT RESPONSE											
Rise Time (Notes 1, 5, 6 & 8)	+25°C		30	60		30	60		30	60	ns
Overshoot (Notes 1, 5, 7 & 8)	+25°C		25	40		25	40		25	40	%
* Slew Rate (Notes 1, 5, 8 & 12)	+25°C	±4	±7		±4	±7		±4	±7		V/µs
Settling Time (Notes 1, 5, 8 & 12)	+25°C		1.5			1.5			1.5		µs
POWER SUPPLY CHARACTERISTICS											
* Supply Current	+25°C		3.0	3.7		3.0	4.0		3.0	4.0	mA
* Power Supply Rejection Ratio (Note 9)	Full	80	90		74	90		74	90		dB

TEST CONDITIONS

NOTES: 1. R_L = 2K
     2. $V_{CM} = \pm 10V$
     3. $V_O < 90mV$
     4. $V_O = \pm 10V$
     5. C_L = 100pF
     6. $V_O = \pm 200mV$

7. $V_O = \pm 200mV$
8. See Transient response test circuits and waveforms
9. $\Delta VS = \pm 5V$

10. This parameter value guaranteed by design calculations. ·
11. Full power bandwidth guaranteed by slew rate measurement. FPBW = S.R./2π V_{peak}.
12. $V_{OUT} = \pm 5V$

*100% Tested For DASH 8

V+ = 15VDC, V- = 15VDC, T$_A$ = 25°C UNLESS OTHERWISE STATED.

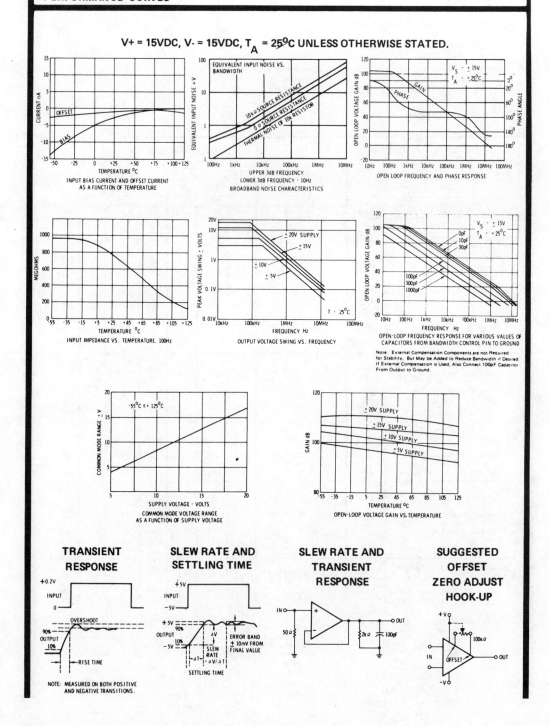

INPUT BIAS CURRENT AND OFFSET CURRENT
AS A FUNCTION OF TEMPERATURE

EQUIVALENT INPUT NOISE VS. BANDWIDTH

UPPER 3dB FREQUENCY
LOWER 3dB FREQUENCY - 10Hz
BROADBAND NOISE CHARACTERISTICS

OPEN LOOP FREQUENCY AND PHASE RESPONSE

INPUT IMPEDANCE VS. TEMPERATURE, 100Hz

OUTPUT VOLTAGE SWING VS. FREQUENCY

OPEN-LOOP FREQUENCY RESPONSE FOR VARIOUS VALUES OF
CAPACITORS FROM BANDWIDTH CONTROL PIN TO GROUND

Note: External Compensation Components are not Required
for Stability, But May be Added to Reduce Bandwidth if Desired.
If External Compensation is Used, Also Connect 100pF Capacitor
From Output to Ground.

COMMON MODE VOLTAGE RANGE
AS A FUNCTION OF SUPPLY VOLTAGE

OPEN-LOOP VOLTAGE GAIN VS. TEMPERATURE

TRANSIENT RESPONSE

NOTE: MEASURED ON BOTH POSITIVE
AND NEGATIVE TRANSITIONS.

SLEW RATE AND SETTLING TIME

SLEW RATE AND TRANSIENT RESPONSE

SUGGESTED OFFSET ZERO ADJUST HOOK-UP

PHOTO-CURRENT TO VOLTAGE CONVERTER

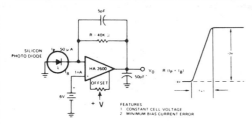

FEATURES
1 CONSTANT CELL VOLTAGE
2 MINIMUM BIAS CURRENT ERROR

SAMPLE – AND – HOLD

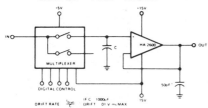

REFERENCE VOLTAGE AMPLIFIER

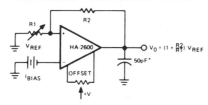

FEATURES
1 MINIMUM BIAS CURRENT IN REFERENCE CELL
2 SHORT CIRCUIT PROTECTION

VOLTAGE FOLLOWER

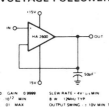

*A small load capacitance is recommended in all applications where practical to prevent possible high frequency oscillations resulting from external wiring parasitics. Capacitance up to 100pF has negligible effect on the bandwidth or slew rate.

INPUT OFFSET VOLTAGE — That voltage which must be applied between the input terminals through two equal resistances to force the output voltage to zero.

INPUT OFFSET CURRENT — The difference in the currents into the two input terminals when the output is at zero voltage.

INPUT BIAS CURRENT — The average of the currents flowing into the input terminals when the output is at zero voltage.

INPUT COMMON MODE VOLTAGE — The average referred to ground of the voltages at the two input terminals.

COMMON MODE RANGE — The range of voltages which is exceeded at either input terminal will cause the amplifier to cease operating.

COMMON MODE REJECTION RATIO — The ratio of a specified range of input common mode voltage to the peak-to-peak change in input offset voltage over this range.

OUTPUT VOLTAGE SWING — The peak symmetrical output voltage swing, referred to ground, that can be obtained without clipping.

INPUT RESISTANCE — The ratio of the change in input voltage to the change in input current.

OUTPUT RESISTANCE — The ratio of the change in output voltage to the change in output current.

VOLTAGE GAIN — The ratio of the change in output voltage to the change in input voltage producing it.

BANDWIDTH — The frequency at which the voltage gain is 3dB below its low frequency value.

UNITY GAIN BANDWIDTH — The frequency at which the voltage gain of the amplifier is unity.

POWER SUPPLY REJECTION RATIO — The ratio of the change in input offset voltage to the change in power supply voltage producing it.

TRANSIENT RESPONSE — The closed loop step function response of the amplifier under small signal conditions.

PHASE MARGIN — $(180^\circ - (\phi_1 - \phi_2))$ where ϕ_1 is the phase shift at the frequency where the absolute magnitude of gain is unity ϕ_2 is the phase shift at a frequency much lower than the open loop bandwidth.

SLEW RATE (Rate Limiting) — The rate at which the output will move between full scale stops, measured in terms of volts per unit time. This limit to an ideal step function response is due to the non-linear behavior in an amplifier due to its limited ability to produce large, rapid changes in output voltage (slewing) . . . . restricting it to rates of change of voltage lower than might be predicted by observing the small signal frequency response.

SETTLING TIME — Time required for output waveform to remain within 0.1% of final value.

SILECT† TRANSISTORS‡

- Ideal for Low-Level Amplifier Applications
- Rugged One-Piece Construction with In-Line Leads or Standard TO-18 100-mil Pin-Circle Configuration
- Recommended for Complementary Use with 2N4058 thru 2N4062, A5T4058 thru A5T4062, or A8T4058 thru A8T4062

mechanical data

These transistors are encapsulated in a plastic compound specifically designed for this purpose, using a highly mechanized process developed by Texas Instruments. The case will withstand soldering temperatures without deformation. These devices exhibit stable characteristics under high-humidity conditions and are capable of meeting MIL-STD-202C, Method 106B. The transistors are insensitive to light.

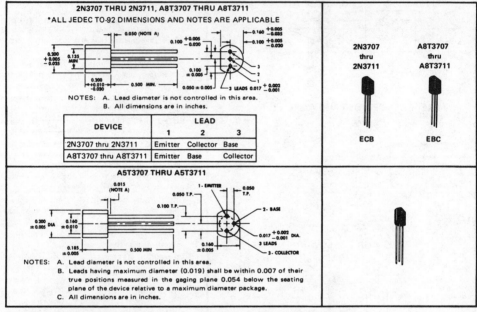

2N3707 THRU 2N3711, A8T3707 THRU A8T3711

*ALL JEDEC TO-92 DIMENSIONS AND NOTES ARE APPLICABLE

NOTES: A. Lead diameter is not controlled in this area.
B. All dimensions are in inches.

DEVICE	LEAD		
	1	2	3
2N3707 thru 2N3711	Emitter	Collector	Base
A8T3707 thru A8T3711	Emitter	Base	Collector

2N3707 thru 2N3711 — ECB

A8T3707 thru A8T3711 — EBC

A5T3707 THRU A5T3711

NOTES: A. Lead diameter is not controlled in this area.
B. Leads having maximum diameter (0.019) shall be within 0.007 of their true positions measured in the gaging plane 0.054 below the seating plane of the device relative to a maximum diameter package.
C. All dimensions are in inches.

absolute maximum ratings at 25°C free-air temperature (unless otherwise noted)

Collector-Base Voltage	30 V*
Collector-Emitter Voltage (See Note 1)	30 V*
Emitter-Base Voltage	6 V*
Continuous Collector Current	30 mA*
Continuous Device Dissipation at (or below) 25°C Free-Air Temperature (See Note 2)	{625 mW§ / 360 mW*
Storage Temperature Range	−65°C to 150°C*
Lead Temperature 1/16 Inch from Case for 10 Seconds	260°C*

NOTES: 1. This value applies when the base-emitter diode is open-circuited.
2. Derate the 625-mW rating linearly to 150°C free-air temperature at the rate of 5 mW/°C. Derate the 360-mW (JEDEC registered) rating linearly to 150°C free-air temperature at the rate of 2.88 mW/°C.

*The asterisk identifies JEDEC registered data for the 2N3707 through 2N3711 only. This data sheet contains all applicable registered data in effect at the time of publication.
†Trademark of Texas Instruments
‡U.S. Patent No. 3,439,238
§Texas Instruments guarantees this value in addition to the JEDEC registered value which is also shown.

USES CHIP N21

From "The Transistor and Diode Data Book for Design Engineers" © 1973, Texas Instruments Incorporated.

240

*electrical characteristics at 25°C free-air temperature

PARAMETER		TEST CONDITIONS	2N3707 A5T3707 A8T3707		2N3708 A5T3708 A8T3708		2N3709 A5T3709 A8T3709		2N3710 A5T3710 A8T3710		2N3711 A5T3711 A8T3711		UNIT
			MIN	MAX	MIN	MAX	MIN	MAX	MIN	MAX	MIN	MAX	
$V_{(BR)CEO}$	Collector-Emitter Breakdown Voltage	I_C = 1 mA, I_B = 0	30		30		30		30		30		V
I_{CBO}	Collector Cutoff Current	V_{CB} = 20 V, I_E = 0		100		100		100		100		100	nA
I_{EBO}	Emitter Cutoff Current	V_{EB} = 6 V, I_C = 0		100		100		100		100		100	nA
h_{FE}	Static Forward Current Transfer Ratio	V_{CE} = 5 V, I_C = 100 μA	100	400									
		V_{CE} = 5 V, I_C = 1 mA			45	660	45	165	90	330	180	660	
V_{BE}	Base-Emitter Voltage	V_{CE} = 5 V, I_C = 1 mA	0.5	1	0.5	1	0.5	1	0.5	1	0.5	1	V
$V_{CE(sat)}$	Collector-Emitter Saturation Voltage	I_B = 0.5 mA, I_C = 10 mA		1		1		1		1		1	V
h_{fe}	Small-Signal Common-Emitter Forward Current Transfer Ratio	V_{CE} = 5 V, I_C = 100 μA, f = 1 kHz	100	550									
		V_{CE} = 5 V, I_C = 1 mA, f = 1 kHz			45	800	45	250	90	450	180	800	

*operating characteristics at 25°C free-air temperature

PARAMETER		TEST CONDITIONS			2N3707, A5T3707, A8T3707			UNIT
					MIN	TYP	MAX	
$\overline{F}$	Average Noise Figure	V_{CE} = 5 V, I_C = 100 μA, Noise Bandwidth = 15.7 kHz,	R_G = 5 kΩ, See Note 3			1.9	5	dB

NOTE 3: Average Noise Figure is measured in an amplifier with response down 3 dB at 10 Hz and 10 kHz and a high-frequency rolloff of 6 dB/octave.

*The asterisk identifies JEDEC registered data for 2N3707 through 2N3711 only.

THERMAL INFORMATION

DISSIPATION DERATING CURVE

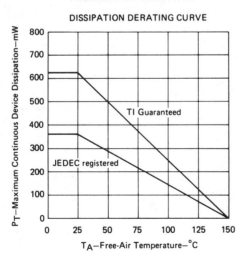

FIGURE 1

TYPICAL CHARACTERISTICS

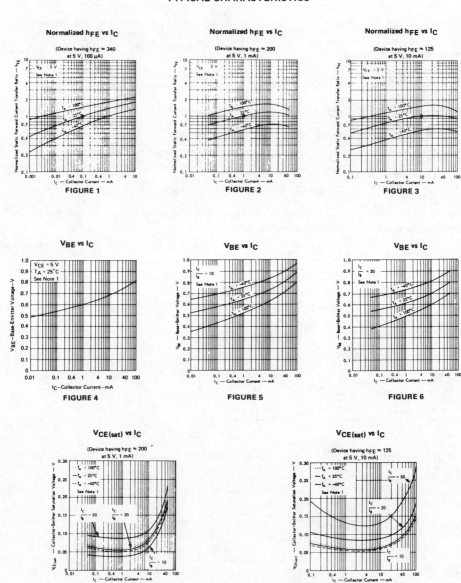

FIGURE 1

FIGURE 2

FIGURE 3

FIGURE 4

FIGURE 5

FIGURE 6

FIGURE 7

FIGURE 8

NOTE 1: These parameters were measured using pulse techniques. $t_w = 300\ \mu s$, duty cycle $\leqslant 2\%$.

TYPICAL CHARACTERISTICS

Normalized h_{ie}, h_{fe}, h_{re}, h_{oe} vs I_C

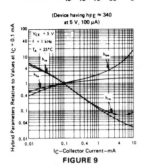

(Device having $h_{FE} \approx 340$ at 5 V, 100 μA)

FIGURE 9

Normalized h_{ie}, h_{fe}, h_{re}, h_{oe} ve I_C

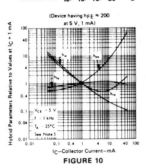

(Device having $h_{FE} \approx 200$ at 5 V, 1 mA)

FIGURE 10

Normalized h_{ie}, h_{fe}, h_{re}, h_{oe} vs I_C

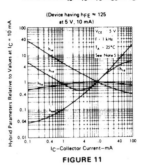

(Device having $h_{FE} \approx 125$ at 5 V, 10 mA)

FIGURE 11

Normalized h_{ie}, h_{fe}, h_{re}, h_{oe} vs V_{CE}

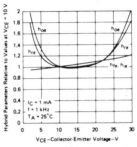

FIGURE 12

f_T vs I_C

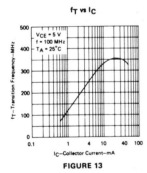

FIGURE 13

C_{cb} vs V_{CB}

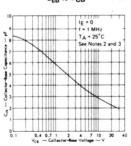

FIGURE 14

C_{eb} vs V_{EB}

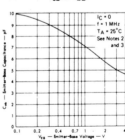

FIGURE 15

NOTES: 2. Capacitance measurements were made using chips mounted in *Silect*† packages.
3. C_{cb} and C_{eb} measurements employ a three-terminal capacitance bridge incorporating a guard circuit. The third electrode (emitter or collector, respectively) is connected to the guard terminal of the bridge.
5. To obtain reproducible results, these parameters were measured with bias conditions applied for less than five seconds.

TYPES 2N4123, 2N4124, A5T4123, A5T4124
N-P-N SILICON TRANSISTORS

BULLETIN NO. DL-S 7311471, NOVEMBER 1971—REVISED MARCH 1973

SILECT† TRANSISTORS‡
DESIGNED FOR GENERAL PURPOSE SATURATED SWITCHING AND AMPLIFIER APPLICATIONS

- **For Complementary Use with P-N-P Types 2N4125, 2N4126, A5T4125, and A5T4126**
- **Rugged One-Piece Construction with In-Line Leads or Standard TO-18 100-mil Pin-Circle Configuration**

mechanical data

These transistors are encapsulated in a plastic compound specifically designed for this purpose, using a highly mechanized process developed by Texas Instruments. This case will withstand soldering temperatures without deformation. These devices exhibit stable characteristics under high-humidity conditions and are capable of meeting MIL-STD-202C, Method 106B. The transistors are insensitive to light.

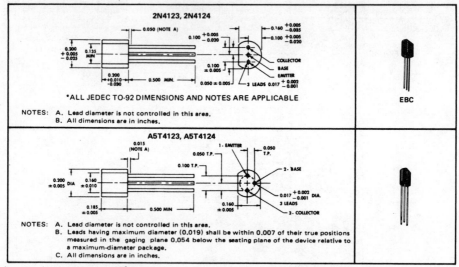

2N4123, 2N4124

*ALL JEDEC TO-92 DIMENSIONS AND NOTES ARE APPLICABLE

EBC

NOTES: A. Lead diameter is not controlled in this area.
B. All dimensions are in inches.

A5T4123, A5T4124

NOTES: A. Lead diameter is not controlled in this area.
B. Leads having maximum diameter (0.019) shall be within 0.007 of their true positions measured in the gaging plane 0.054 below the seating plane of the device relative to a maximum-diameter package.
C. All dimensions are in inches.

absolute maximum ratings at 25°C free-air temperature (unless otherwise noted)

	2N4123 A5T4123	2N4124 A5T4124
Collector-Base Voltage	40 V*	30 V*
Collector-Emitter Voltage (See Note 1)	30 V*	25 V*
Emitter-Base Voltage	5 V*	5 V*
Continuous Collector Current	←——— 200 mA* ———→	
Continuous Device Dissipation at (or below) 25°C Free-Air Temperature (See Note 2)	←— {625 mW§ / 310 mW*}	
Storage Temperature Range	{−65°C to 150°C§ / −55°C to 135°C*}	
Lead Temperature 1/16 Inch from Case for 60 Seconds	←——— 230°C* ———→	

NOTES: 1. These values apply between 10 µA and 200 mA collector current when the base-emitter diode is open-circuited.
2. Derate the 625-mW rating linearly to 150°C free-air temperature at the rate of 5 mW/°C. Derate the 310-mW (JEDEC registered) rating linearly to 135°C free-air temperature at the rate of 2.81 mW/°C.

*The asterisk identifies JEDEC registered data for the 2N4123 and 2N4124 only. This data sheet contains all applicable registered data in effect at the time of publication.
†Trademark of Texas Instruments
‡U.S. Patent No. 3,439,238
§Texas Instruments guarantees these values in addition to the JEDEC registered values which are also shown.

USES CHIP N14

From "The Transistor and Diode Data Book for Design Engineers" © 1973, Texas Instruments Incorporated.

*electrical characteristics at 25°C free-air temperature

	PARAMETER	TEST CONDITIONS		2N4123 A5T4123		2N4124 A5T4124		UNIT
				MIN	MAX	MIN	MAX	
$V_{(BR)CBO}$	Collector-Base Breakdown Voltage	$I_C = 10\,\mu A$, $I_E = 0$		40		30		V
$V_{(BR)CEO}$	Collector-Emitter Breakdown Voltage	$I_C = 1\,mA$, $I_B = 0$,	See Note 3	30		25		V
$V_{(BR)EBO}$	Emitter-Base Breakdown Voltage	$I_E = 10\,\mu A$, $I_C = 0$		5		5		V
I_{CBO}	Collector Cutoff Current	$V_{CB} = 20\,V$, $I_E = 0$			50		50	nA
I_{EBO}	Emitter Cutoff Current	$V_{EB} = 3\,V$, $I_C = 0$			50		50	nA
h_{FE}	Static Forward Current Transfer Ratio	$V_{CE} = 1\,V$, $I_C = 2\,mA$	See Note 3	50	150	120	360	
		$V_{CE} = 1\,V$, $I_C = 50\,mA$		25		60		
V_{BE}	Base-Emitter Voltage	$I_B = 5\,mA$, $I_C = 50\,mA$,	See Note 3		0.95		0.95	V
$V_{CE(sat)}$	Collector-Emitter Saturation Voltage	$I_B = 5\,mA$, $I_C = 50\,mA$,	See Note 3		0.3		0.3	V
h_{fe}	Small-Signal Common-Emitter Forward Current Transfer Ratio	$V_{CE} = 10\,V$, $I_C = 2\,mA$, $f = 1\,kHz$		50	200	120	480	
$\|h_{fe}\|$	Small-Signal Common-Emitter Forward Current Transfer Ratio	$V_{CE} = 20\,V$, $I_C = 10\,mA$, $f = 100\,MHz$		2.5		3		
f_T	Transition Frequency	$V_{CE} = 20\,V$, $I_C = 10\,mA$, See Note 4		250		300		MHz
C_{obo}	Common-Base Open-Circuit Output Capacitance	$V_{CB} = 5\,V$, $I_E = 0$, $f = 100\,kHz$			4		4	pF
C_{ibo}	Common-Base Open-Circuit Input Capacitance	$V_{EB} = 0.5\,V$, $I_C = 0$, $f = 100\,kHz$			8		8	pF

NOTES: 3. These parameters must be measured using pulse techniques. $t_w = 300\,\mu s$, duty cycle $\leqslant 2\%$.
     4. To obtain f_T, the $\|h_{fe}\|$ response is extrapolated at the rate of $-6\,dB$ per octave from $f = 100\,MHz$ to the frequency at which $\|h_{fe}\| = 1$.

*operating characteristics at 25°C free-air temperature

	PARAMETER	TEST CONDITIONS		2N4123 A5T4123		2N4124 A5T4124		UNIT
				MIN	MAX	MIN	MAX	
$\overline{NF}$	Average Noise Figure	$V_{CE} = 5\,V$, $I_C = 100\,\mu A$, $R_G = 1\,k\Omega$, Noise Bandwidth = 15.7 kHz, See Note 5			6		5	dB

NOTE 5: Average noise figure is measured in an amplifier with response down 3 dB at 10 Hz and 10 kHz and a high-frequency rolloff of 6 dB/octave.
*The asterisk identifies JEDEC registered data for the 2N4123 and 2N4124 only.

switching characteristics at 25°C free-air temperature

	PARAMETER	TEST CONDITIONS†	TYP	UNIT
t_d	Delay Time	$I_C = 10\,mA$, $I_{B(1)} = 1\,mA$, $V_{BE(off)} = -0.5\,V$,	14	ns
t_r	Rise Time	$R_L = 275\,\Omega$, See Figure 1	8	ns
t_s	Storage Time	$I_C = 10\,mA$, $I_{B(1)} = 1\,mA$, $I_{B(2)} = -1\,mA$,	22	ns
t_f	Fall Time	$R_L = 275\,\Omega$, See Figure 2	10	ns

†Voltage and current values shown are nominal; exact values vary slightly with transistor parameters. Nominal base current for delay and rise times is calculated using the minimum value of V_{BE}. Nominal base currents for storage and fall times are calculated using the maximum value of V_{BE}.

PARAMETER MEASUREMENT INFORMATION

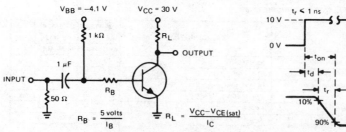

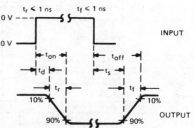

$$R_B = \frac{5 \text{ volts}}{I_B}$$

$$R_L = \frac{V_{CC} - V_{CE(sat)}}{I_C}$$

TEST CIRCUIT

VOLTAGE WAVEFORMS

(See Notes a and b)

NOTES: a. The input waveforms are supplied by a generator with the following characteristics: Z_{out} = 50 Ω; for measuring t_d and t_r, t_w ≈ 200 ns, duty cycle ≤ 2%; for measuring t_s and t_f, t_w ≈ 10 µs, duty cycle ≤ 2%.

b. Waveforms are monitored on an oscilloscope with the following characteristics: t_r ≤ 1 ns, R_{in} ≥ 100 kΩ, C_{in} ≤ 7 pF.

FIGURE 1—SWITCHING TIMES

TYPICAL CHARACTERISTICS

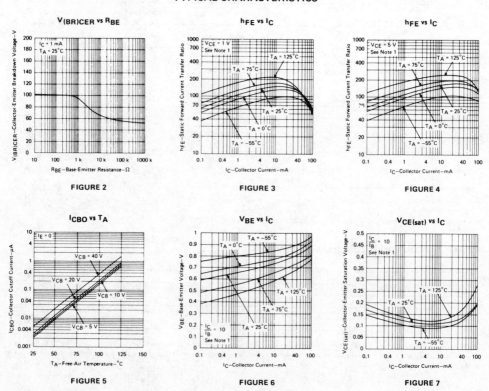

NOTE 1: These parameters were measured using pulse techniques. t_w = 300 µs, duty cycle ≤ 2%.

TYPICAL CHARACTERISTICS

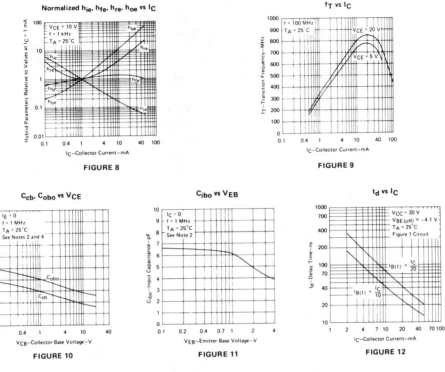

FIGURE 8

FIGURE 9

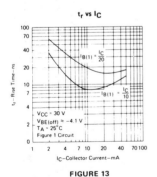

FIGURE 10

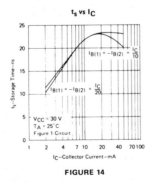

FIGURE 11

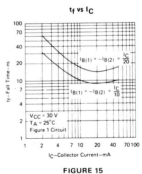

FIGURE 12

FIGURE 13

FIGURE 14

FIGURE 15

NOTES: 1. These parameters were measured using pulse techniques. t_w = 300 μs, duty cycle ≤ 2%.
2. Capacitance measurements were made using chips mounted in *Silect* packages.
4. C_{cb} measurement employs a three-terminal capacitance bridge incorporating a guard circuit. The emitter is connected to the guard terminal of the bridge. C_{obo} measurement is made with the third terminal floating.

N-CHANNEL SILECT† FIELD-EFFECT TRANSISTORS‡
FOR VHF AMPLIFIER AND MIXER APPLICATIONS

- **High Power Gain . . . 10 dB Min at 400 MHz**
- **High Transconductance . . . 4000 μmho Min at 400 MHz (2N5245, 2N5247)**
- **Low C_{rss} . . . 1 pF Max**
- **High $|y_{fs}|/C_{iss}$ Ratio (High-Frequency Figure-of-Merit)**
- **Drain and Gate Leads Separated for High Maximum Stable Gain**
- **Cross-Modulation Minimized by Square-Law Transfer Characteristic**
- **For Use in VHF Amplifiers in FM, TV, and Mobile Communications Equipment**

mechanical data

These transistors are encapsulated in a plastic compound specifically designed for this purpose, using a highly mechanized process developed by Texas Instruments. The case will withstand soldering temperatures without deformation. These devices exhibit stable characteristics under high-humidity conditions and are capable of meeting MIL-STD-202C, Method 106B. The transistors are insensitive to light.

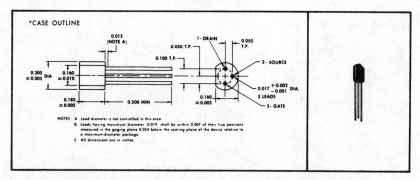

*absolute maximum ratings at 25°C free-air temperature (unless otherwise noted)

Drain-Gate Voltage	30 V
Reverse Gate-Source Voltage	−30 V
Continuous Forward Gate Current	50 mA
Continuous Device Dissipation at (or below) 25°C Free-Air Temperature (See Note 1)	360 mW
Continuous Device Dissipation at (or below) 25°C Lead Temperature (See Note 2)	500 mW
Storage Temperature Range	−65°C to 150°C
Lead Temperature 1/16 Inch from Case for 10 Seconds	260°C

NOTES: 1. Derate linearly to 150°C free-air temperature at the rate of 2.88 mW/°C.

2. Derate linearly to 150°C lead temperature at the rate of 4 mW/°C. Lead temperature is measured on the gate lead 1/16 inch from the case.

*Indicates JEDEC registered data

†Trademark of Texas Instruments

‡U.S. Patent No. 3,439,238

USES CHIP JN53

From "The Transistor and Diode Data Book for Design Engineers" © 1973, Texas Instruments Incorporated.

*electrical characteristics at 25°C free-air temperature (unless otherwise noted)

PARAMETER		TEST CONDITIONS	2N5245 MIN	2N5245 MAX	2N5246 MIN	2N5246 MAX	2N5247 MIN	2N5247 MAX	UNIT
$V_{(BR)GSS}$	Gate-Source Breakdown Voltage	$I_G = -1\ \mu A,\ V_{DS} = 0$	-30		-30		-30		V
I_{GSS}	Gate Reverse Current	$V_{GS} = -20\ V,\ V_{DS} = 0$		-1		-1		-1	nA
		$V_{GS} = -20\ V,\ V_{DS} = 0,\quad T_A = 100°C$		-0.5		-0.5		-0.5	μA
$V_{GS(off)}$	Gate-Source Cutoff Voltage	$V_{DS} = 15\ V,\quad I_D = 10\ nA$	-1	-6	-0.5	-4	-1.5	-8	V
I_{DSS}	Zero-Gate-Voltage Drain Current	$V_{DS} = 15\ V,\quad V_{GS} = 0,\quad$ See Note 3	5	15	1.5	7	8	24	mA
$|y_{fs}|$	Small-Signal Common-Source Forward Transfer Admittance	$V_{DS} = 15\ V,\quad V_{GS} = 0,\quad f = 1\ kHz$	4.5	7.5	3	6	4.5	8	mmho
$|y_{os}|$	Small-Signal Common-Source Output Admittance	$V_{DS} = 15\ V,\quad V_{GS} = 0,\quad f = 1\ kHz$		0.05		0.05		0.07	mmho
C_{iss}	Common-Source Short-Circuit Input Capacitance	$V_{DS} = 15\ V,$		4.5		4.5		4.5	pF
C_{rss}	Common-Source Short-Circuit Reverse Transfer Capacitance	$V_{GS} = 0,$ $f = 1\ MHz$		1		1		1	pF
$Re(y_{is})$	Small-Signal Common-Source Input Conductance	$V_{DS} = 15\ V,$		0.1		0.1		0.1	mmho
$Im(y_{is})$	Small-Signal Common-Source Input Susceptance	$V_{GS} = 0,$		3		3		3	mmho
$Re(y_{os})$	Small-Signal Common-Source Output Conductance	$f = 100\ MHz$		0.075		0.075		0.1	mmho
$Im(y_{os})$	Small-Signal Common-Source Output Susceptance			1		1		1	mmho
$Re(y_{is})$	Small-Signal Common-Source Input Conductance			1		1		1	mmho
$Im(y_{is})$	Small-Signal Common-Source Input Susceptance	$V_{DS} = 15\ V,$		12		12		12	mmho
$Re(y_{fs})$	Small-Signal Common-Source Forward Transfer Conductance	$V_{GS} = 0,$	4		2.5		4		mmho
$Re(y_{os})$	Small-Signal Common-Source Output Conductance	$f = 400\ MHz$		0.1		0.1		0.15	mmho
$Im(y_{os})$	Small-Signal Common-Source Output Susceptance			4		4		4	mmho

NOTE 3: This parameter must be measured using pulse techniques. $t_p = 100\ ms$, duty cycle $\leq 10\%$.

*operating characteristics at 25°C free-air temperature

PARAMETER		TEST CONDITIONS	2N5245 MIN	2N5245 MAX	UNIT
G_{ps}	Small-Signal Common-Source Neutralized Insertion Power Gain	$V_{DS} = 15\ V,\quad I_D = 5\ mA,\quad f = 100\ MHz,$ $R_G' = 1\ k\Omega,\quad$ See Figure 1	18		dB
		$V_{DS} = 15\ V,\quad I_D = 5\ mA,\quad f = 400\ MHz,$ $R_G' = 1\ k\Omega,\quad$ See Figure 1	10		
NF	Spot Noise Figure	$V_{DS} = 15\ V,\quad I_D = 5\ mA,\quad f = 100\ MHz,$ $R_G' = 1\ k\Omega,\quad$ See Figure 1		2	dB
		$V_{DS} = 15\ V,\quad I_D = 5\ mA,\quad f = 400\ MHz,$ $R_G' = 1\ k\Omega,\quad$ See Figure 1		4	

*Indicates JEDEC registered data

TYPES TIP29, TIP29A, TIP29B, TIP29C
BULLETIN NO. DL-S 7011371, OCTOBER 1970
REPLACES BULLETIN NO. DL-S 6810954, JULY 1968

FOR POWER-AMPLIFIER AND HIGH-SPEED-SWITCHING APPLICATIONS
DESIGNED FOR COMPLEMENTARY USE WITH TIP30, TIP30A, TIP30B, TIP30C

- 30 W at 25°C Case Temperature
- 1 A Rated Collector Current
- Min f_T of 3 MHz at 10 V, 200 mA

mechanical data

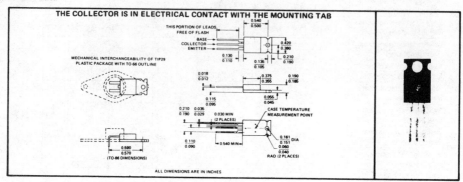

THE COLLECTOR IS IN ELECTRICAL CONTACT WITH THE MOUNTING TAB

ALL DIMENSIONS ARE IN INCHES

absolute maximum ratings at 25°C case temperature (unless otherwise noted)

	TIP29	TIP29A	TIP29B	TIP29C
Collector-Base Voltage	40 V	60 V	80 V	100 V
Collector-Emitter Voltage (See Note 1)	40 V	60 V	80 V	100 V
Emitter-Base Voltage	◄———— 5 V ————►			
Continuous Collector Current	◄———— 1 A ————►			
Peak Collector Current (See Note 2)	◄———— 3 A ————►			
Continuous Base Current	◄———— 0.4 A ————►			
Safe Operating Region at (or below) 25°C Case Temperature	◄———— See Figure 5 ————►			
Continuous Device Dissipation at (or below) 25°C Case Temperature (See Note 3)	◄———— 30 W ————►			
Continuous Device Dissipation at (or below) 25°C Free-Air Temperature (See Note 4)	◄———— 2 W ————►			
Unclamped Inductive Load Energy (See Note 5)	◄———— 32 mJ ————►			
Operating Collector Junction Temperature Range	◄———— −65°C to 150°C ————►			
Storage Temperature Range	◄———— −65°C to 150°C ————►			
Lead Temperature 1/8 Inch from Case for 10 Seconds	◄———— 260°C ————►			

NOTES: 1. This value applies when the base-emitter diode is open-circuited.
2. This value applies for $t_w \leqslant 0.3$ ms, duty cycle $\leqslant 10\%$.
3. Derate linearly to 150°C case temperature at the rate of 0.24 W/°C.
4. Derate linearly to 150°C free-air temperature at the rate of 16 mW/°C.
5. This rating is based on the capability of the transistor to operate safely in the circuit of Figure 2. L = 20 mH, R_{BB2} = 100 Ω, V_{BB2} = 0 V, R_S = 0.1 Ω, V_{CC} = 10 V. Energy ≈ $I_C^2 L/2$.

From "The Power Semiconductor Data Book for Design Engineers" © 1980,
Texas Instruments Incorporated.

electrical characteristics at 25°C case temperature

PARAMETER		TEST CONDITIONS		TIP29		TIP29A		TIP29B		TIP29C		UNIT
				MIN	MAX	MIN	MAX	MIN	MAX	MIN	MAX	
$V_{(BR)CEO}$	Collector-Emitter Breakdown Voltage	$I_C = 30$ mA, $\quad I_B = 0$, See Note 6		40		60		80		100		V
I_{CEO}	Collector Cutoff Current	$V_{CE} = 30$ V, $\quad I_B = 0$			0.3		0.3					mA
		$V_{CE} = 60$ V, $\quad I_B = 0$							0.3		0.3	
I_{CES}	Collector Cutoff Current	$V_{CE} = 40$ V, $\quad V_{BE} = 0$			0.2							mA
		$V_{CE} = 60$ V, $\quad V_{BE} = 0$					0.2					
		$V_{CE} = 80$ V, $\quad V_{BE} = 0$							0.2			
		$V_{CE} = 100$ V, $\quad V_{BE} = 0$									0.2	
I_{EBO}	Emitter Cutoff Current	$V_{EB} = 5$ V, $\quad I_C = 0$			1		1		1		1	mA
h_{FE}	Static Forward Current Transfer Ratio	$V_{CE} = 4$ V, $\quad I_C = 0.2$ A, See Notes 6 and 7		40		40		40		40		
		$V_{CE} = 4$ V, $\quad I_C = 1$ A, See Notes 6 and 7		15	75	15	75	15	75	15	75	
V_{BE}	Base-Emitter Voltage	$V_{CE} = 4$ V, $\quad I_C = 1$ A, See Notes 6 and 7			1.3		1.3		1.3		1.3	V
$V_{CE(sat)}$	Collector-Emitter Saturation Voltage	$I_B = 125$ mA, $\quad I_C = 1$ A, See Notes 6 and 7			0.7		0.7		0.7		0.7	V
h_{fe}	Small-Signal Common-Emitter Forward Current Transfer Ratio	$V_{CE} = 10$ V, $\quad I_C = 0.2$ A, $f = 1$ kHz		20		20		20		20		
$|h_{fe}|$	Small-Signal Common-Emitter Forward Current Transfer Ratio	$V_{CE} = 10$ V, $\quad I_C = 0.2$ A, $f = 1$ MHz		3		3		3		3		

NOTES: 6. These parameters must be measured using pulse techniques. $t_w = 300$ μs, duty cycle ≤ 2%.
       7. These parameters are measured with voltage-sensing contacts separate from the current-carrying contacts.

thermal characteristics

PARAMETER		MAX	UNIT
$R_{\theta JC}$	Junction-to-Case Thermal Resistance	4.17	°C/W
$R_{\theta JA}$	Junction-to-Free-Air Thermal Resistance	62.5	

switching characteristics at 25°C case temperature

PARAMETER		TEST CONDITIONS[†]			TYP	UNIT
t_{on}	Turn-On Time	$I_C = 1$ A,	$I_{B(1)} = 100$ mA, $\quad I_{B(2)} = -100$ mA,		0.5	μs
t_{off}	Turn-Off Time	$V_{BE(off)} = -4.3$ V, $\quad R_L = 30$ Ω,		See Figure 1	2	

[†]Voltage and current values shown are nominal; exact values vary slightly with transistor parameters.

TYPES TIP29, TIP29A, TIP29B, TIP29C
N-P-N SINGLE-DIFFUSED MESA SILICON POWER TRANSISTORS

TYPICAL CHARACTERISTICS
STATIC FORWARD CURRENT TRANSFER RATIO
vs
COLLECTOR CURRENT

THERMAL INFORMATION

DISSIPATION DERATING CURVE

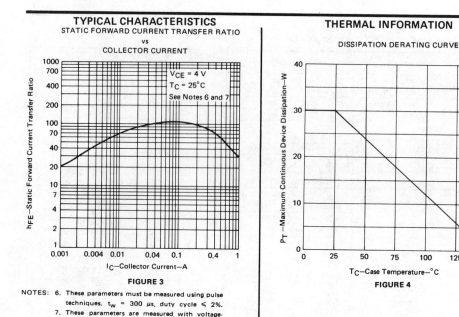

FIGURE 3

FIGURE 4

NOTES: 6. These parameters must be measured using pulse
techniques. t_W = 300 µs, duty cycle ≤ 2%.
7. These parameters are measured with voltage-
sensing contacts separate from the current-
carrying contacts.

MAXIMUM SAFE OPERATING REGION

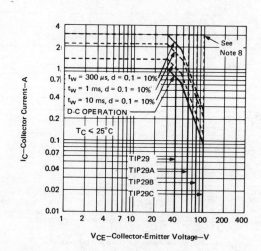

FIGURE 5

NOTE 8: This combination of maximum voltage and current may be achieved only when switching from saturation to cutoff with a clamped
inductive load.

DESIGNED FOR LOW-NOISE PREAMPLIFIER APPLICATIONS ESPECIALLY HYDROPHONES, IR SENSORS, AND PARTICLE DETECTORS

- Low V_n . . . 5 nV/$\sqrt{Hz}$ Max at 10 Hz (2N6451, 2N6453)
- High $|y_{fs}|$ . . . 20 mmho Min (2N6453, 2N6454)
- Low I_{GSS} . . . 100 pA Max (2N6451, 2N6453)

*mechanical data

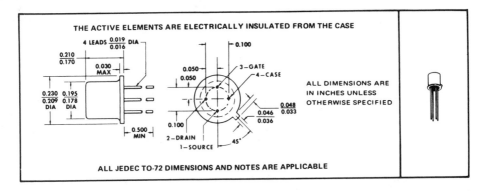

THE ACTIVE ELEMENTS ARE ELECTRICALLY INSULATED FROM THE CASE

ALL DIMENSIONS ARE IN INCHES UNLESS OTHERWISE SPECIFIED

ALL JEDEC TO-72 DIMENSIONS AND NOTES ARE APPLICABLE

*absolute maximum ratings at 25°C free-air temperature (unless otherwise noted)

	2N6451 2N6453	2N6452 2N6454
Drain-Gate Voltage	20 V	25 V
Reverse Gate-Source Voltage	−20 V	−25 V
Continuous Forward Gate Current	◄——10 mA——►	
Continuous Device Dissipation at (or below) 25°C Free-Air Temperature (See Note 1)	◄——360 mW——◄	
Storage Temperature Range	−65°C to 200°C	
Lead Temperature 1/16 Inch from Case for 10 Seconds	◄—— 300°C ——►	

NOTE 1: Derate linearly to 175°C free-air temperature at the rate of 2.4 mW/°C.

*JEDEC registered data. This data sheet contains all applicable registered data in effect at the time of publication.

USES CHIP JN55

TYPES 2N6451 THRU 2N6454
N-CHANNEL JUNCTION GATE FIELD-EFFECT TRANSISTORS

*electrical characteristics at 25°C free-air temperature (unless otherwise noted)

PARAMETER		TEST CONDITIONS†	2N6451		2N6452		2N6453		2N6454		UNIT
			MIN	MAX	MIN	MAX	MIN	MAX	MIN	MAX	
$V_{(BR)GSS}$	Gate-Source Breakdown Voltage	$I_G = -1\ \mu A$, $V_{DS} = 0$	−20		−25		−20		−25		V
I_{GSS}	Gate Reverse Current	$V_{GS} = -10\ V$, $V_{DS} = 0$		−0.1				−0.1			nA
		$V_{GS} = -15\ V$, $V_{DS} = 0$				−0.5				−0.5	
		$V_{GS} = -10\ V$, $V_{DS} = 0$ $T_A = 125°C$		−0.2				−0.2			µA
		$V_{GS} = -15\ V$, $V_{DS} = 0$				−1				−1	
$V_{GS(off)}$	Gate-Source Cutoff Voltage	$V_{DS} = 10\ V$, $I_D = 0.5\ nA$	−0.5	−3.5	−0.5	−3.5	−0.75	−5	−0.75	−5	V
I_{DSS}	Zero-Gate-Voltage Drain Current	$V_{DS} = 10\ V$, $V_{GS} = 0$, See Note 2	5	20	5	20	15	50	15	50	mA
$|y_{fs}|$	Small-Signal Common-Source Forward Transfer Admittance	$V_{DS} = 10\ V$, $I_D = 5\ mA$, $f = 1\ kHz$	15	30	15	30					mmho
		$V_{DS} = 10\ V$, $I_D = 15\ mA$, $f = 1\ kHz$, See Note 3					20	40	20	40	
$|y_{os}|$	Small-Signal Common-Source Output Admittance	$V_{DS} = 10\ V$, $I_D = 5\ mA$, $f = 1\ kHz$		50		50					µmho
		$V_{DS} = 10\ V$, $I_D = 15\ mA$, $f = 1\ kHz$, See Note 3						100		100	
C_{iss}	Common-Source Short-Circuit Input Capacitance	$V_{DS} = 10\ V$, $I_D = 5\ mA$, $f = 1\ MHz$		25		25					pF
		$V_{DS} = 10\ V$, $I_D = 15\ mA$, $f = 1\ MHz$, See Note 3						25		25	
C_{rss}	Common-Source Short-Circuit Reverse Transfer Capacitance	$V_{DS} = 10\ V$, $I_D = 5\ mA$, $f = 1\ MHz$		5		5					pF
		$V_{DS} = 10\ V$, $I_D = 15\ mA$, $f = 1\ MHz$, See Note 3						5		5	

*operating characteristics at 25°C free-air temperature

PARAMETER		TEST CONDITIONS†	2N6451		2N6452		2N6453		2N6454		UNIT
			MIN	MAX	MIN	MAX	MIN	MAX	MIN	MAX	
F	Common-Source Spot Noise Figure	$V_{DS} = 10\ V$, $I_D = 5\ mA$, $R_G = 10\ k\Omega$, $f = 10\ Hz$		1.5		2.5		1.5		2.5	dB
V_n	Equivalent Input Noise Voltage	$V_{DS} = 10\ V$, $I_D = 5\ mA$, $f = 10\ Hz$		5		10		5		10	nV/√Hz
		$V_{DS} = 10\ V$, $I_D = 5\ mA$, $f = 1\ kHz$		3		8		3		8	

*JEDEC registered data
†The fourth lead (case) is connected to the source for all measurements.
NOTES: 2. This parameter must be measured using pulse techniques. $t_w = 300\ \mu s$, duty cycle ≤ 2%.
    3. To obtain repeatable results, this parameter must be measured with bias conditions applied for less than five seconds.

2N**5088** (SILICON)
2N**5089**

NPN SILICON ANNULAR TRANSISTORS

. . . NPN silicon annular transistors designed for low-level, low- noise amplifier applications.

- High Gain at Low Current —
  2N5088 — 300 min at 100 μAdc
  2N5089 — 400 min at 100 μAdc
- Low 100 μAdc Noise Figure —
  1.2 dB typ at 100 Hz
  1.0 dB typ at 1.0 kHz
- Excellent Gain Linearity from 20 μAdc to 2.0 mAdc
  (See Figure 9)

NPN SILICON AMPLIFIER TRANSISTORS

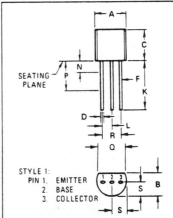

*MAXIMUM RATINGS

Rating	Symbol	2N5088	2N5089	Unit
Collector-Emitter Voltage	V_{CEO}	30	25	Vdc
Collector-Base Voltage	V_{CB}	35	30	Vdc
Emitter-Base Voltage	V_{EB}	4.5		Vdc
Collector Current — Continuous	I_C	50		mAdc
Total Power Dissipation @ T_A = 25°C Derate above 25°C	P_D	350 2.8		mW mW/°C
Total Power Dissipation @ T_C = 25°C Derate above 25°C	P_D	1.0 8.0		Watt mW/°C
Operating and Storage Junction Temperature Range	T_J, T_{stg}	-55 to +150		°C

THERMAL CHARACTERISTICS

Characteristic	Symbol	Max	Unit
Thermal Resistance, Junction to Ambient	$R_{\theta JA}$(1)	357	°C/W
Thermal Resistance, Junction to Case	$R_{\theta JC}$	125	°C/W

*Indicates JEDEC Registered Data
(1) $R_{\theta JA}$ is measured with the device soldered into a typical printed circuit board.

STYLE 1:
PIN 1. EMITTER
     2. BASE
     3. COLLECTOR

DIM	MILLIMETERS		INCHES	
	MIN	MAX	MIN	MAX
A	4.450	5.200	0.175	0.205
B	3.180	4.190	0.125	0.165
C	4.320	5.330	0.170	0.210
D	0.407	0.533	0.016	0.021
F	0.407	0.482	0.016	0.019
K	12.700	.-	0.500	-
L	1.150	1.390	0.045	0.055
N	–	1.270	–	0.050
P	6.350	-	0.250	-
Q	3.430	–	0.135	–
R	2.410	2.670	0.095	0.105
S	2.030	2.670	0.080	0.105

CASE 29-02
TO-92

ELECTRICAL CHARACTERISTICS (T$_A$ = 25°C unless otherwise noted)

Characteristic		Symbol	Min	Max	Unit
OFF CHARACTERISTICS					
Collector-Emitter Breakdown Voltage(1) (I$_C$ = 1.0 mAdc, I$_B$ = 0) 2N5088 2N5089		BV$_{CEO}$	30 25	- -	Vdc
Collector-Base Breakdown Voltage (I$_C$ = 100 μAdc, I$_E$ = 0) 2N5088 2N5089		BV$_{CBO}$	35 30	- -	Vdc
Collector Cutoff Current (V$_{CB}$ = 20 Vdc, I$_E$ = 0) 2N5088 (V$_{CB}$ = 15 Vdc, I$_E$ = 0) 2N5089		I$_{CBO}$	- -	50 50	nAdc
Emitter Cutoff Current (V$_{EB(off)}$ = 3.0 Vdc, I$_C$ = 0) (V$_{EB(off)}$ = 4.5 Vdc, I$_C$ = 0)		I$_{EBO}$	- -	50 100	nAdc
ON CHARACTERISTICS					
DC Current Gain (I$_C$ = 100 mAdc, V$_{CE}$ = 5.0 Vdc) 2N5088 2N5089 (I$_C$ = 1.0 mAdc, V$_{CE}$ = 5.0 Vdc) 2N5088 2N5089 (I$_C$ = 10 mAdc, V$_{CE}$ = 5.0 Vdc)(1) 2N5088 2N5089		h$_{FE}$	300 400 350 450 300 400	900 1200 - - - -	
Collector-Emitter Saturation Voltage (I$_C$ = 10 mAdc, I$_B$ = 1.0 mAdc)		V$_{CE(sat)}$	-	0.5	Vdc
Base-Emitter On Voltage (I$_C$ = 10 mAdc, V$_{CE}$ = 5.0 Vdc)(1)		V$_{BE(on)}$	-	0.8	Vdc
DYNAMIC CHARACTERISTICS					
Current-Gain – Bandwidth Product (I$_C$ = 500 μAdc, V$_{CE}$ = 5.0 Vdc, f = 20 MHz		f$_T$	50	-	MHz
Collector-Base Capacitance (V$_{CB}$ = 5.0 Vdc, I$_E$ = 0, f = 100 kHz)		C$_{cb}$	-	4.0	pF
Emitter-Base Capacitance (V$_{BE}$ = 0.5 Vdc, I$_C$ = 0, f = 100 kHz)		C$_{eb}$	-	10	pF
Small-Signal Current Gain (I$_C$ = 1.0 mAdc, V$_{CE}$ = 5.0 Vdc, f = 1.0 kHz) 2N5088 2N5089		h$_{fe}$	350 450	1400 1800	-
Noise Figure (I$_C$ = 100 μAdc, V$_{CE}$ = 5.0 Vdc, R$_S$ = 10 k ohms, 2N5088 f = 10 Hz to 15.7 kHz 2N5089		NF	- -	3.0 2.0	dB

* Indicates JEDEC Registered Data
(1) Pulse Test: Pulse Width ≤ 300 μs, Duty Cycle ≤ 2.0%.

NOISE FIGURE
V$_{CE}$ = 5.0 Vdc, T$_A$ = 25°C

FIGURE 1 – FREQUENCY EFFECTS   **FIGURE 2 – SOURCE RESISTANCE EFFECTS**

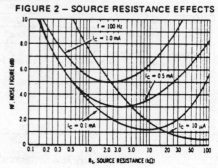

FIGURE 3 – DC CURRENT GAIN

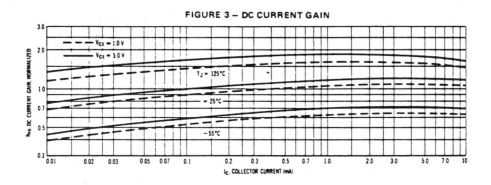

FIGURE 4 – COLLECTOR SATURATION REGION

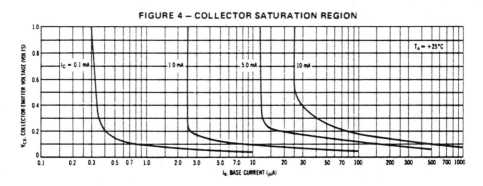

FIGURE 5 – CURRENT-GAIN–
BANDWIDTH PRODUCT

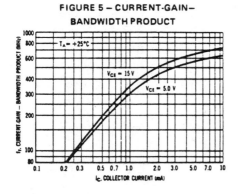

FIGURE 6 – CAPACITANCE

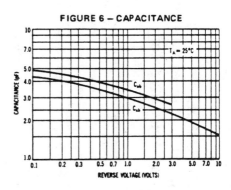

CD4007A Types

COS/MOS
Dual Complementary
Pair Plus Inverter

The RCA-CD4007A types are comprised of three n-channel and three p-channel enhancement-type MOS transistors. The transistor elements are accessible through the package terminals to provide a convenient means for constructing the various typical circuits as shown in Fig. 2.

More complex functions are possible using multiple packages. Numbers shown in parentheses indicate terminals that are connected together to form the various configurations listed.

The CD4007-A Series types are supplied in 14-lead hermetic dual-in-line ceramic packages (D,F, and Y suffixes), 14-lead dual-in-line plastic packages (E suffix), 14-lead ceramic flat packages (K suffix), and in chip form (H suffix).

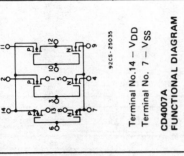

92CS-25035

Terminal No.14 — V_{DD}
Terminal No. 7 — V_{SS}

**CD4007A
FUNCTIONAL DIAGRAM**

Features:

- Medium-speed operation. . . . . $t_{PHL} = t_{PLH} = 20$ ns (typ.) at $C_L = 15$ pF, + $V_{DD} = 10$ V
- Low "high" and "low" output impedance. . . . . 500 Ω (typ.) at $V_{DD} - V_{SS} = 10$ V
- Quiescent current specified to 15 V
- Maximum input leakage current of 1 μA at 15 V (full package-temperature range)
- 1-V noise margin (full package-temperature range)

Applications:

- Extremely high-input impedance amplifiers
- Shapers
- Inverters
- Threshold detector
- Linear amplifiers

MAXIMUM RATINGS, Absolute-Maximum Values:

STORAGE-TEMPERATURE RANGE (T_{stg})	−65 to +150°C
OPERATING-TEMPERATURE RANGE (T_A)	
PACKAGE TYPES D, F, K, H	−55 to +125°C
PACKAGE TYPES E, Y	−40 to +85°C
DC SUPPLY-VOLTAGE RANGE, (V_{DD}) (Voltages referenced to V_{SS} Terminal)	−0.5 to +15 V
POWER DISSIPATION PER PACKAGE (P_D)	
FOR $T_A = -40$ to +60°C (PACKAGE TYPES E, Y)	500 mW
FOR $T_A = +60$ to +85°C (PACKAGE TYPES E, Y) Derate Linearly at 12 mW/°C to 200 mW	
FOR $T_A = -55$ to +100°C (PACKAGE TYPES D, F, K)	500 mW
FOR $T_A = +100$ to +125°C (PACKAGE TYPES D, F, K) Derate Linearly at 12 mW/°C to 200 mW	
DEVICE DISSIPATION PER OUTPUT TRANSISTOR	
FOR $T_A = $ FULL PACKAGE-TEMPERATURE RANGE (ALL PACKAGE TYPES)	100 mW
INPUT VOLTAGE RANGE, ALL INPUTS	−0.5 to V_{DD} +0.5 V
LEAD TEMPERATURE (DURING SOLDERING)	
At distance 1/16 ± 1/32 inch (1.59 ± 0.79 mm) from case for 10 s max	+265°C

From "RCA Integrated Circuits" © 1976, The Radio Corporation of America.

RECOMMENDED OPERATING CONDITIONS

For maximum reliability, nominal operating conditions should be selected so that operation is always within the following ranges:

CHARACTERISTIC	LIMITS				UNITS
	D,F,K,H Packages		E,Y Packages		
	Min.	Max.	Min.	Max.	
Supply-Voltage Range (For T$_A$ = Full Package Temperature Range)	3	12	3	12	V

DYNAMIC ELECTRICAL CHARACTERISTICS at T$_A$ = 25°C, Input t$_r$, t$_f$ = 20 ns, C$_L$ = 15 pF, R$_L$ = 200 kΩ

CHARACTERISTIC	TEST CONDITIONS		LIMITS						UNITS
		V$_{DD}$ (V)	D,F,K,H Packages			E,Y Packages			
			Min.	Typ.	Max.	Min.	Typ.	Max.	
Propagation Delay Time; tPLH, tPHL		5	–	35	60	–	35	75	ns
		10	–	20	40	–	20	50	
Transition Time; tTHL, tTLH		5	–	50	75	–	50	100	ns
		10	–	30	40	–	30	50	
Average Input Capacitance, C$_I$	Any Input		–	5	–	–	5	–	pF

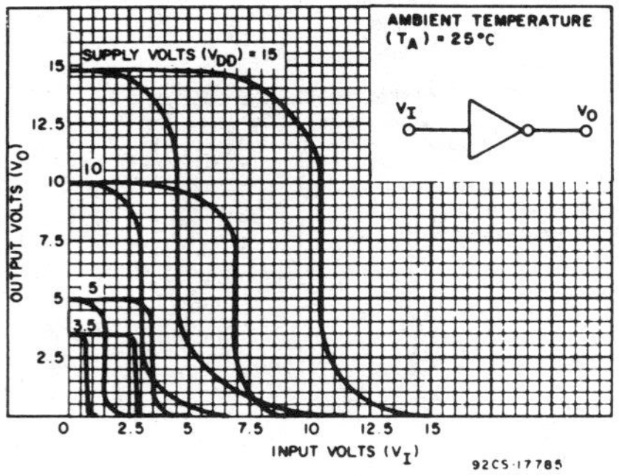

Fig.1 — Minimum and maximum voltage-transfer characteristics for inverter.

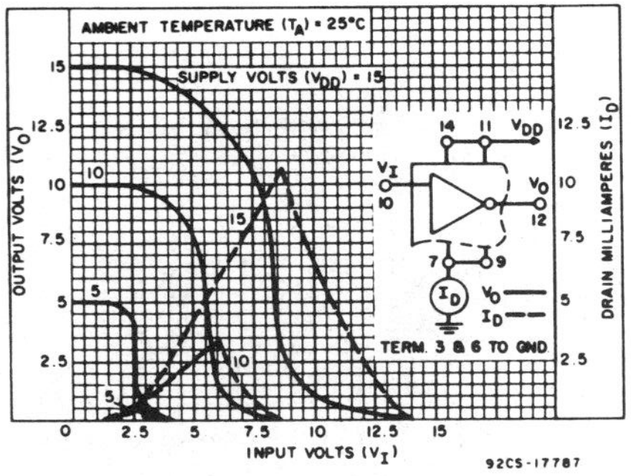

Fig.2 — Typical current and voltage-transfer characteristics for inverter.

CD4007A Types

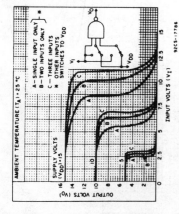

Fig.3 – Typical voltage-transfer characteristics for NAND gate.

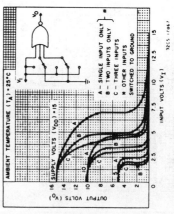

Fig.4 – Typical voltage-transfer characteristics for NOR gate.

STATIC ELECTRICAL CHARACTERISTICS

Characteristic	VO (V)	VIN (V)	VDD (V)	D,K,F,H Packages −55	+25 Typ.	+25 Limit	+125	E,Y Packages −40	+25 Typ.	+25 Limit	+85	Units
Quiescent Device Current: IL Max.	–	5	5	0.05	0.001	0.05	3	0.5	0.005	0.5	15	μA
	–	10	10	0.1	0.001	0.1	6	1	0.005	1	30	
	–	–	15	2	0.02	2	40	50	0.5	50	500	
Output Voltage Low Level V_{OL}	–	5	5	0 Typ.; 0.05 Max.								V
	–	10	10	0 Typ.; 0.05 Max.								
High Level V_{OH}	–	0	5	4.95 Min.; 5 Typ.								
	–	0	10	9.95 Min.; 10 Typ.								
Noise Immunity: Inputs Low V_{NL}	3.6	–	5	1.5 Min.; 2.25 Typ.								V
	7.2	–	10	3 Min.; 4.5 Typ.								
Inputs High V_{NH}	1.4	–	5	1.5 Min.; 2.25 Typ.								
	2.8	–	10	3 Min.; 4.5 Typ.								
Noise Margin: Inputs Low V_{NML}	4.5	–	5	1 Min.								V
	9	–	10	1 Min.								
Inputs High V_{NMH}	0.5	–	5	1 Min.								
	1	–	10	1 Min.								
Output Drive Current: N-Channel (Sink) IDN Min.	0.5	$V_I=$	5	0.75	1	0.6	0.4	0.35	1	0.3	0.24	mA
	0.5	V_{DD}	10	1.6	2.5	1.3	0.95	1.2	2.5	1	0.8	
P-Channel (Source) IDP Min.	4.5	$V_I=$	5	−1.75	−4	−1.4	−1	−1.3	−4	−1.1	−0.9	
	9.5	V_{DD}	10	−1.35	−2.5	−1.1	−0.75	−0.65	−2.5	−0.55	−0.45	
Input Leakage Current: I_{IL}, I_{IH}	Any Input	–	15	$\pm10^{-5}$ Typ., ±1 Max.								μA

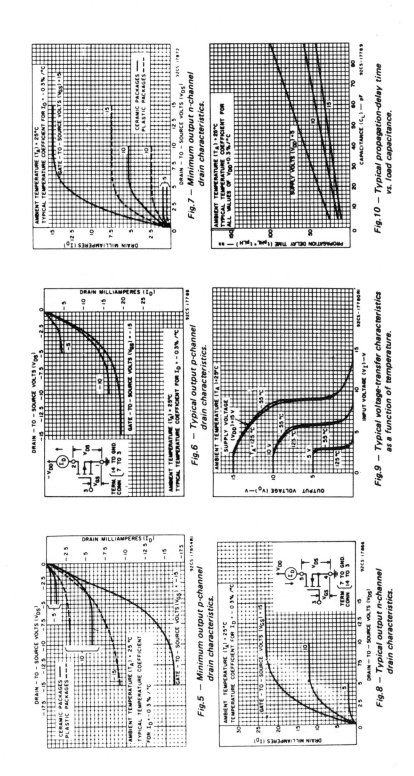

Fig.7 — Minimum output n-channel drain characteristics.

Fig.10 — Typical propagation-delay time vs. load capacitance.

Fig.6 — Typical output p-channel drain characteristics.

Fig.9 — Typical voltage-transfer characteristics as a function of temperature.

Fig.5 — Minimum output p-channel drain characteristics.

Fig.8 — Typical output n-channel drain characteristics.

261

12-21-83